THE GREEN BUBBLE

FOR GREEN ENERGY TO BE TRULY SUSTAINABLE
IT MUST BE COMMERCIALLY SUSTAINABLE

THE GREEN BUBBLE

FOR GREEN ENERGY TO BE TRULY SUSTAINABLE
IT MUST BE COMMERCIALLY SUSTAINABLE

BY PER WIMMER

in collaboration with
EDWARD RUSSELL-WALLING

LONDON   MONTERREY
MADRID   SHANGHAI
MEXICO CITY   BOGOTA
NEW YORK   BUENOS AIRES
BARCELONA   SAN FRANCISCO

Published by
LID Publishing Ltd
One Adam Street, London WC2N 6LE

31 West 34th Street, Suite 7004,
New York, NY 10001, US

info@lidpublishing.com
www.lidpublishing.com

A member of:

BPR
Business Publishers Roundtable

www.businesspublishersroundtable.com

Printed in Great Britain by TJ International
ISBN: 978-1-907794-89-6
Cover and page design: Laura Hawkins

CONTENTS

CHAPTER 1:

GETTING OUR FIX

At lunchtime on 30 July 2012, three out of India's five electricity grids crashed. The two-day blackout that followed upset the lives, to varying degrees, of an estimated 700 million people. Hundreds of trains stopped without warning, many in the middle of nowhere. Failed traffic lights caused major traffic jams, bringing cities to a standstill. Water supplies were interrupted, surgical operations and cremations were halted, and miners were stranded below ground. When demand for power outweighs a country's ability to provide it, the lights go out. And they could start going out in more developed countries before long.

India has a slightly better excuse than the developed world for losing the electricity plot – it is an emerging economy whose rapid growth has outgrown its creaking old infrastructure. If the lights go out in the West, it will be because we became so infatuated with green energy that we let it distort our grip on reality. And the results could be extreme. Although previous power cuts have provoked the occasional riot, the discomfited Indians were fairly restrained in their response to the Great Blackout. And yet mayhem is never far from the surface. In the summer of 1977, a series of lightning strikes knocked out the electricity supply to most of New York City for more than 24 hours. In the almost total darkness that resulted, civilization appeared to evaporate. More than 1,600 shops were looted, more than 1,000 fires broke out, and an outburst of general violence and disorder prompted nearly 4,000 arrests. It was said that the city's birth rate spiked nine months later. This turned out to be false, but people had no trouble believing it, because it made perfect intuitive sense. Without electricity, with no light or cooking facilities, no television or public transport, and with anarchy in the streets, our nocturnal options are reduced to those of our Neolithic ancestors.

The fact is that we are hopelessly addicted to energy. It is the thin glue that holds our 'civilized' world together and life without it would be unthinkable. We need it to provide the life-or-death necessities of food, clothing, and shelter, as well as the comforts we have come to regard as necessities – lighting, heat, phones, computers, televisions, and all those other gadgets. We require energy in one form or another to travel more than a few kilometres by land, air, or, for the most part, sea. This is an addiction for which there is no cure and, like most addicts, our habit demands bigger and bigger fixes. Unlike street junkies, however, we need to plan many years in advance exactly where those fixes will come from.

That would be challenging enough in itself, but we have an additional problem. Our drug, or at least the way we produce it, has unpleasant side effects. The emission of greenhouse gases and other pollutants associated with fossil fuels is clearly undesirable, and replacing them with cleaner, more sustainable fuels is a rational goal. Whether it is sensible or even possible to try to achieve that goal in a compressed time frame, and at any price, is a different matter. The idea of promoting green energy while cutting emissions has such broad and instant popular appeal that it has shot up the political agenda in most developed countries. This political Bubble, which has diverted large sums of public money into renewable energy, has in turn created a financial Bubble, as business snaps up the 'free' money on offer to launch otherwise unviable green energy projects. Both Bubbles are unsustainable and when they burst, as they must, they will not only create financial havoc but may also set back the green energy cause for many years.

The fundamental, urgent question is how do we keep the lights on in 2030, or 2040, or 2050 while reducing carbon emissions in a meaningful way? As taxpayers, we have a crucial part to play in determining the answer, since it is our money that politicians are spending on one specific vision of a green energy future. Politics drives energy policy as much, or even more, than economics, and the two do not always pull in the same direction. Dependent as we are on finite supplies of dirty fossil fuels, often under the control of foreign, and not always friendly, governments, we want our energy to be clean and sustainable. But we also want it to be *there* and our aspiration must be tempered by what is practical and possible. Politicians are inclined to court short-term popularity at the expense of the hard choices necessary for a long-term good such as energy security. Yet there is no perfect energy option. The best we can do is to make the least-worst choice. This book will argue that there are more effective ways of spending our money to achieve the least-worst outcome.

What do I know about it, and why should you pay any attention to what I think? I am an investment banker, specializing in mining, oil and gas, and alternative energy companies. A Dane, I now live and work in London. After a career with a number of leading financial companies, including Goldman Sachs in New York, I set up my own investment bank, Wimmer Financial, with a network of corporate clients, institutional investors, and business partners in other geographies. I have a separate asset management company, Wimmer Family Office, which runs investment strategies on behalf of high-net-worth individuals and families. Wimmer Space is the umbrella company for my three planned trips to space and other adventures, including the world's first tandem skydive over Mount Everest (in October 2008). It also includes our charity activities, publishing and TV documentaries, as well as public and motivational speaking. Every day I work with energy and resources businesses of all kinds, helping them to raise finance and to plan their futures. So this is a milieu I know a little about, and it is my experience of it that has shaped the thoughts in this book. Before I get into the thrust of my argument, however, let me briefly paint a picture of the energy world that these companies, and ultimately all of us, inhabit.

Since the start of the Industrial Revolution, world energy consumption has increased exponentially – which is to say faster and faster. At one time, its growth was driven by the industrialization of the Western world.

In future, it will be driven more by rising populations and living standards in Asia and Africa. More disposable nappies are now sold in Nigeria each year than in the whole of the European Union, which illustrates how the balance of consumption – and, ultimately, energy use – is shifting. As more people in China buy cars and fridges or climb onto planes, so the world's demand for energy will keep on growing.

If energy growth keeps accelerating, the rate at which we have been able to switch from one dominant source of fuel to another has been much, much slower. The Industrial Revolution, which began halfway through the 18th century, was powered mainly by coal. Even so, it was many years before coal overtook wood (or 'biomass', as scientists would say) as the leading energy source. That only happened more than a century later, about 1880, just as oil was starting to appear. Today, oil is our largest energy source, its growth driven by the spread of the automobile and the internal combustion engine. And yet oil's share of the total did not overtake coal until 1960. As we need to understand when contemplating the future, sweeping shifts in fuel usage do not happen overnight, even if we want them to.

Some years ago, an American bioengineer named Hewitt Crane coined an unsettling new term that put our voracious consumption into perspective. Crane was an early IT pioneer who developed the world's first all-magnetic computer and who subsequently worked on optical character and handwriting recognition. Late in his life, he turned to what he realized was a very serious problem: given the speed at which our energy needs are growing, how can we keep supplying them in the future? Surveying all the different fuel sources and their Babel-like units of measurement, he was struck by how difficult it was for non-specialists merely to grasp the scale of the problem – all those mind-numbing tons of coal, barrels of oil, and cubic feet of gas, producing joules, calories, and megawatt-hours, in their millions and billions and trillions. He decided that, if we were to make critical choices about our energy future – which is essentially the subject of this book – we needed common, understandable terms in which to discuss it.

So Crane distilled the bewildering confusion of the energy lexicon into one simple measurement – the 'cubic mile of oil' or CMO. This is exactly what it says, a cube of oil, one mile high, one mile long, and one mile wide. As a unit, it describes our consumption of the real oil that is pumped out of the ground. It also measures the consumption of other energy sources by expressing their thermal energy content in terms of their oil equivalent. So one CMO of coal is the amount of coal that, when burnt, releases the same amount of thermal energy as one CMO of oil. One cubic mile of oil is roughly four cubic kilometres of oil. Expressed in the huge numbers that Crane was trying to avoid, it is also (very) approximately 1 trillion gallons, 4 trillion litres, or 26 billion barrels.

Crane died in 2008, so did not live to see the 2010 publication of *A Cubic Mile of Oil*, the book he co-wrote with colleagues Edwin Kinderman and Ripudaman Malhotra. It examines the options for averting what it calls "the looming global energy crisis" and we shall return to it later. Right now, in their terms of measurement, the world consumes roughly three cubic miles of oil every year.[1] The bad news, according to Crane et al, is that within less than 40 years, by 2050, we will be consuming six cubic miles of oil, and then only if we succeed in our best efforts at energy conservation. If, as is quite possible, we do not, they estimate we will need nine cubic miles of oil. Where will it come from?

One cubic mile or one third of our current consumption is actual oil, with the remaining two thirds coming from other sources such as coal, gas, nuclear, and hydro (see table page 17). It is worth noting that in 2009 all the other 'alternative', 'green', or 'renewable' energy sources combined, such as solar, wind, geothermal, wave, and tide energy but excluding hydro and biomass, made up less than 1% of the total, according to the International Energy Agency (IEA).[2] In its central outlook scenario, the IEA predicts that, by 2035, this share will still not be much more than 4%.

Each of these energy sources has its own distinct character, its own pluses and minuses in terms of cost, energy content and efficiency, abundance and sustainability of supply, technological challenge, and social acceptability. One very important variable, which gets a lot of air time in debates about renewable energy, is 'load factor'. This measures how much electricity a generator type actually produces as a percentage of its total capacity – which is what it would produce if it ran for 24 hours a day and 365 days a year. Whereas a nuclear plant can grind on day and night to achieve load factors of 90% or more, the intermittency of the wind means that onshore wind turbines typically produce electricity less than one third of the time. All of these features will influence the part each plays in our energy supply in years to come, even as our taxes are spent – as, in principle, they should be – in an attempt to mould the desired outcome. Since this is ultimately a political choice, we need to decide what that outcome should be, within the limits of what is technologically possible and what we can reasonably afford.

We have to start by assessing our options, and so the rest of this chapter reviews the fuels and technologies we rely on at present, weighing their advantages and disadvantages. In Chapter 2, we take a critical look at alternative, renewable energy sources for power generation, and Chapter 3 examines the technologies that aspire to replace liquid fossil fuels in the wider world of transport. Alternatives need subsidies, so Chapter 4 looks at how green energy businesses are assembled, at the people who run them, and at how the finances work. Free money usually attracts too many people, however, and already we can see a Bubble forming in the green energy business. Chapter 5 describes this phenomenon and drills deeper into some of the problems attached to renewable energy. Perhaps the greenest approach to energy is to use less of it, so Chapter 6 reports on progress in the energy efficiency market, including smart grids, and light emitting diodes (LEDs). Chapter 7 compares European and US energy strategies, and finds Europe wanting. Chapter 8 shows how the drive for more renewables is starting to go horribly wrong. In the final chapter, all of this evidence is reviewed and I recommend what I believe to be our least-worst option. Let us begin with our traditional energy sources.

COAL

Oil is the biggest single source of our world's energy. But one third of our energy – one cubic mile of oil – is used to generate electricity, and the dominant fuel for electricity generation is coal. It has persuasive advantages, alongside some rather unpleasant disadvantages. Coal has been with us for thousands of years, with traces found in Bronze Age funeral pyres dating from the third millennium BC. It came into its own after James Watt modified the steam engine – the 'external' combustion engine – so that it could provide a rotary motion, as wind and water mills had done for millennia. That was in the latter half of the 18th century and it gave wings to the Industrial Revolution. Burning coal provided the heat that boiled the water to produce the steam. It still does that today, producing steam to turn turbines in coal-fired power stations, although the coal is now milled into powder so that it burns more quickly. 'Supercritical' plants operate at higher temperatures and pressures and produce more energy from less coal, with lower emissions. The latest 'ultra-supercritical' plants do the same but more so – they have only been made possible by new super alloys that can withstand these higher temperatures.

Coal has a thermal content more than twice that of wood, which it gradually replaced as the dominant source of energy. Like other fossil fuels, it is effectively frozen solar energy, absorbed by plants millions of years ago and then buried before it could be released through decay. There are ascending qualities of coal, depending on the nature of the original vegetation and on how long and at what depths it has been buried (longer and deeper means more energy content). Lignite or brown coal is at the low end of the quality spectrum, containing the least thermal energy, rising through sub-bituminous coal, thermal or steam coal, metallurgical coal (used to make iron and steel) to anthracite, which has the fewest impurities and highest calorific content, and is widely used as domestic fuel. The lower-quality coals from lignite to thermal coal are used in power generation, and account for 41% of the world's electricity, followed by gas (21%), hydro (16%), nuclear (13%), and oil (5%) [IEA 2009].

Coal's virtues are that it is plentiful, cheap, and easy to mine. It is found in at least 70 countries around the world, with the biggest reserves in the US, Russia, China, and India. China, which overtook the US as the world's largest energy consumer in 2009, is now the world's biggest producer and biggest consumer of coal. Nearly 80% of China's power generation is coal-fired, and it has been building new coal-burning power stations at the extraordinary rate of about two a week, although an old plant must now be closed for every one that opens. India is

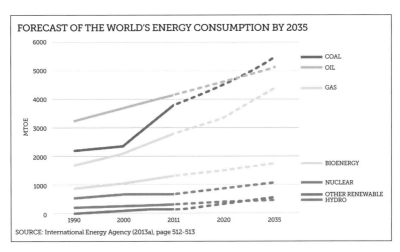

FORECAST OF THE WORLD'S ENERGY CONSUMPTION BY 2035

SOURCE: International Energy Agency (2013a), page 512-513

FORECAST OF WORLD ENERGY CONSUMPTION IN PERCENT PER SOURCE					
ENERGY TYPE	1990	2000	2011	2020	2035
COAL	25	24	26	29	29
OIL	37	36	31	30	27
GAS	19	21	21	22	23
NUCLEAR	6	7	5	6	5
HYDROPOWER	2	2	2	2	3
BIOMASS	10	10	10	10	9
OTHER RENEWABLE	0	1	1	2	3

expected to replace the US as the world's second-largest coal consumer (and its largest seaborne coal importer) by 2017.[3]

Recently, the world ratio of coal reserves to coal production has been falling, making some wonder whether we are approaching the 'peak coal' moment, after which production will go into irreversible decline. As a fossil fuel, coal is a necessarily finite resource and when it has gone, it has gone. It seems unlikely, however, that supplies are anywhere near serious depletion. New discoveries continue to be made and new technologies are allowing previously inaccessible deposits to be mined. What is more, improvements in power station design and combustion technology are enabling more energy to be produced from less coal.

Apart from being a finite resource, coal's greatest disadvantage is, of course, the fact that it is dirty. When burned it releases pollutants such as sulphur dioxide and nitrogen oxides, which cause acid rain, and carbon dioxide, a greenhouse gas. Newer 'integrated gasification combined cycle' (IGCC) plants, which gasify the coal before burning it, have significantly lower non-greenhouse gas emissions, although they can cost nearly twice as much to build. Another greenhouse gas, methane, is trapped underground in coal seams and gets released during the mining process. The deeper and older the coal seam, the more methane escapes. However, methane is the main component of natural gas, and to all intents and purposes it *is* natural gas. So if it can be recovered, it is a useful fuel. The process of extracting coal-bed methane, as it is called, involves a lot of water, which must be disposed of and may be contaminated, so some environmentalists refer to coal-bed methane (or 'coal seam gas') as "the evil twin of shale gas".

Methane links in to another negative feature of coal and coal mining, which is that it is dangerous for those who have to mine it, particularly underground. Subterranean methane explosions kill many miners around the world every year, especially in China. More mining operations are now installing methane recovery systems, partly for safety reasons and partly to sell as a fuel. In the case of surface mines, dust, noise, and landscape disfigurement can be disagreeable for those who live nearby.

Cleaner ways of using coal are evolving, although these invariably add to its cost. Some treat the coal beforehand, whereas others gasify the coal and then clean the gas before firing it in a gas turbine. That is a stop on the road to full 'carbon capture and storage' (CCS), which is more talked about than acted upon. CCS is what it says – a process of separating and capturing carbon dioxide before, during, or after combustion of a fossil fuel such as coal.

The gas is then transported by pipeline, and pumped deep underground where it is stored in porous rocks. It can be pumped into oilfields where, by maintaining pressure, it helps the operators to extract more oil, in a process known as 'enhanced oil recovery'. The general idea is that CCS should be fitted to new coal-fired power stations or retro-fitted to old ones, so that we can continue to benefit from coal's virtues while not suffering from its defects. (Which is why some environmentalists deplore it – they say we should we getting rid of coal altogether.) The problem is that CCS is very expensive and reduces efficiency – the amount of energy extracted from the coal. The US Department of Energy reckons that CCS fitted to a new pulverized coal plant may increase the cost of the electricity by up to 80% while reducing efficiency (because it burns extra fuel) by 20% to 30%.[4] There are a handful of commercial-scale CCS projects up and running, but none is attached to a power station. In most cases, they are used in gas fields to reduce the carbon dioxide content of the gas to commercially acceptable levels.

Coal gasification has other applications in transport and chemicals. Using Fischer-Tropsch technology developed in Germany during the 1920s, coal can be gasified and then turned into less-polluting liquid synthetic fuels, including low-sulphur diesel, as well as certain petrochemicals. Oil-from-coal producer Sasol has been doing this in coal-rich South Africa since 1950, although it has yet to catch on elsewhere. Coal-to-fuel could conceivably become more widely attractive if the price of oil continues to rise, as it almost certainly will. 'Underground coal gasification', still in its infancy, uses similar chemistry to gasify coal without removing it from its seam. In essence, it means setting fire to the coal underground, although the process actually heats the coal to the point where, instead of bursting into flame, it separates into syngas. Some of the carbon dioxide can then be extracted, producing a cleaner, cheaper alternative to natural gas. Perhaps not surprisingly, underground coal gasification freaks out environmentalists even more than shale gas and coal-bed methane.

Coal may be dominant, but there are good reasons why we use a mix of other fuels to supply power to the electricity grid, the network that links power generation to the end user. One is simply security in diversity – we do not want to be too dependent on any one fuel. Another is that different types of plant provide different levels of flexibility and cost. In the early days of electricity, standalone generators were built near the homes or factories that consumed their power. The grid, with its long-distance transmission lines, allows power generated anywhere to be tapped by consumers anywhere. Since there is no practical means of storing electricity on this scale, however, the grid must constantly balance supply and demand from all of these different sources, making electricity always available when needed but without too much costly and wasteful generation. Too much power coming into the grid can be as problematic as not enough power. Staying on the line between the two is a tricky feat to pull off and, if the grid operators get it wrong, blackout beckons.

Maintaining this balance economically calls for a mix of generator types. 'Baseload' generators pump out electricity, at or near full, capacity all the time, with only occasional shutdowns for maintenance. 'Load following' generators adjust their output as demand fluctuates during the 24-hour cycle. And 'standby' generation is only switched on when needed, in extreme peak periods of demand or when baseload or load following plants are unable to supply electricity for one reason or another. Because standby and, to some extent, load following require the plants to be fired up and shut down more often, they have higher running costs and shorter lives. In some countries, an additional plant is switched into the grid according to a 'merit order' – those with the lowest marginal cost of production come in first. Marginal costs are variable costs that, in this case, are essentially fuel costs. Since the costs of renewable energy fuel – the sunshine, the wind, the action of the sea – are zero, that puts them at the head of the queue, causing problems for other generators that we will touch on later.

Coal-fired power stations are expensive to build compared with gas-fired plants, although they have a longer technical lifetime, at about 40 years. Today, big plants routinely cost more than $1 billion to build. A proposed 740MW Chilean coal-fired plant now being contested in court has a budget of $1.4 billion[5], for example. That is approximately $1,900 per kilowatt (kW) of capacity. The latest ultra-supercritical plants have a typical investment cost in Europe (lower in China, higher in the US and Japan) of about $2,100 per kW.[6] Newer IGCC designs (that is the integrated gasification combined cycle technology which, as mentioned earlier, gasifies the coal), can cost $2,400 per kW.

Coal plants take about four years to build and can take days to fire up, but they are cheap to run, so older power stations were ideally suited for baseload operations. Newer ones are having to adapt to more load following, particularly as intermittent supplies from renewables such as solar and wind power become more of a permanent feature on the grid. Load factors for coal vary from country to country and plant to plant, but in general, at about 75% to 85%, are the highest of all fuels except nuclear. All in all, coal has historically been the cheapest way to produce bulk electricity. Today, however, the economics of coal are being stretched in some more developed countries by regulatory requirements to control emissions with costly flue cleaning or carbon capture and storage equipment, as well as the need to buy carbon permits equivalent to their emissions. These regulatory costs close the cost gap between coal and cleaner fuels, making the latter more competitive, which is precisely the point of them. Carbon permits are not working as well as they might, because they are so cheap to buy and therefore do not make burning coal as prohibitively expensive as their inventors might have hoped. We will touch on this, the so-called 'price of carbon', again in the next chapter.

Coal may have been cast as power generation's villain, but do not expect to say goodbye to it any time soon. In spite of efforts to relegate it to its smoggy past, coal's share of the global energy mix is actually still rising, and the IEA[7] says that, by 2017, it will "come close to surpassing oil" as the world's top energy source. Between now and then, we will burn another 1.2 billion more tonnes a year of coal, more than the current annual consumption of the US and Russia combined, the agency says. That is because of abundant supplies and the "insatiable" demand for power from emerging markets. "In the absence of a high carbon price, only fierce competition from low-priced gas can effectively reduce coal demand," it concludes. Coal may seem like yesterday's technology, but people need energy, and new coal-fired plants are opening up almost every day. In late 2012, the World Resources Institute identified 1,200 new coal power stations being planned in 59 countries.[8]

OIL AND GAS
Crude oil or petroleum (from the Greek 'petra' and 'oleum' for 'rock' and 'oil') was first used for lighting in the mid-18th century. By the 1880s, realizing that oil had twice the energy content of coal, Admiral of the Fleet Lord Fisher, the great Royal Navy reformer, began arguing for its adoption as fuel in British warships. Replacing coal with oil would increase their range, allow smaller boilers, and do away with the arduous work of loading and stoking coal, freeing up men for more productive tasks. Then, as now, oil packs a lot of punch in a small space. It was the 20th century before the Royal Navy took Fisher's advice and by then oil, or its gasoline derivative, had found a growth market in the shape of the automobile. Even then there was a debate about whether it might be better to power the new vehicles with batteries.

Oil and coal are created in much the same way, except that while coal comes from dead plant matter, oil and gas are produced by millions of dead marine organisms that drifted to the floors of seas and lakes before being covered by sediment. There is no such thing as a subterranean lake of oil. Instead, the oil resides in tiny cavities within porous sedimentary rock, which makes it rather more difficult to extract. At high temperatures oil turns into natural gas, which may be found in the same place as oil or on its own. Major oil sites include the Middle East and North Africa, Siberia, the North Sea, the Gulf of Mexico, West Africa (most notably Nigeria but with regions such as Angola and Southern Sudan becoming more important), Indonesia, and the Caspian Sea. The great unexplored oil territory lies beneath the Arctic Ocean, abutted by Alaska, Canada, Greenland, Norway, and Russia and is now attracting a lot of attention from international oil companies, not to mention governments. The Arctic may hold 13% of the world's estimated undiscovered oil reserves and a whopping 30% of its undiscovered natural gas, according to the *US Geological Survey*.[9]

'Peak oil' is a more pressing debate than 'peak coal', but the day when world oil production begins to head irreversibly downhill is a moving target, as new finds continue to be made. Looking for oil costs increasingly more per barrel discovered, because all the easy deposits have already been found. Oil companies must explore in ever more challenging places, usually at sea, and in ever deeper waters, where costs multiply. The countries that own the oil are taking a larger slice of the proceeds in taxes and royalties, boosting what they call their 'retained economic interests'. Recent upheavals in the Middle East have raised concerns over future oil supplies, which has forced prices upward. So oil economics continue to grow more challenging, even as higher prices justify more costly exploration. Oil companies are still making sizeable finds, in places such as East Africa, offshore Brazil, and in the deep waters off French Guiana. Even in the North Sea, long regarded as in decline and originally supposed to have been sucked dry by 1990, Norway's Statoil recently celebrated the biggest find since the 1980s. And regardless of what has already been found beneath the ocean floor, most of the world's offshore territory has yet to be explored.

Another factor pushing back the date of peak oil is technology, which is allowing oil producers to squeeze more oil out of existing wells or new geologies. Since they often have to leave behind two barrels of oil for every one they extract, this leaves some scope for improvement. If they could increase recovery rates only to one barrel in two it would, self-evidently, double the world's proven oil reserves. Some techniques use heat or injections of gas, chemicals, or water to force more oil to the surface. The latest embryonic idea for squeezing production involves nanotechnology – injecting millions of minute carbon clusters ('nanoreporters') into underground reservoirs, where changes in their chemistry will signal the presence of oil that has been left behind. 'Horizontal' drilling, although considerably more expensive than vertical drilling, can greatly increase production. Horizontal drilling combined with hydraulic fracturing or 'fracking' (explained later), has opened up vast new reserves of 'tight oil' – oil in relatively nonporous rock – in the US, Canada, and quite possibly elsewhere, such as China, Russia, and Argentina.

Unlike a lot of new green energy businesses, where making the sums work can be very sensitive to this subsidy or that interest rate, financing a new oil or gas project is reasonably straightforward. If it costs X to get it out of the ground and you can sell it for Y, and if Y is still bigger than X after all the outstretched hands have had their share, you have got a nice profitable business. Where the wells are situated can be more problematic. Under normal circumstances, you would not get too close to Nigeria, for example. It is very risky. If you want to do deals in Russia, it helps if the parties are politically well-connected.

It would be good if we were less dependent on liquid fuels such as gasoline and fuel oil, since they account for more than a third of all man-made carbon emissions.[10] The bulk of this comes from the world's motor vehicle fleet, which has doubled every 15 years since 1970 and which continues to grow. Meaningful contributions also come from airlines and the shipping industry, although aircraft engines continue to become more efficient and therefore 'cleaner'. If the world's shipping fleet were a country, it would be the sixth largest polluter on earth, according to United Nations statistics. Experimental sails on container ships, to complement rather than replace their engines, suggest that substantial reductions are possible from that quarter. Molten carbonate fuel cells, which make electricity from compressed air and hydrogen-rich syngas (produced from carbons), can already power a ship's auxiliary electrical systems and may one day propel the ship. Cleaner fuel for cars seems a no-brainer, even if the green lobby has now turned its face against certain kinds of biofuels.

Yet, if we are addicted to energy in general, we are seriously hooked on oil and it will be very difficult to get unhooked. It is the most calorific of all the fossil fuels and, like the most pernicious drugs, gives a lot of bang for your buck. With electricity generation, it is possible over time to change fuel sources without undue disruption. The central distribution infrastructure – the electricity grid – remains relatively unaffected. But our transport infrastructure is much more inextricably wedded to petroleum-based fuel, with millions of gasoline- and diesel-powered vehicles and their attendant networks of refuelling and servicing points. We shall explore the possibilities for electric and hydrogen-powered vehicles more fully in Chapter 3, but there remain huge problems with battery technology in areas such as capacity vs. weight, how far you can drive before recharging, and the risks for any private sector initiatives to create a nationwide recharging infrastructure.

If the oil ran out, we would be forced to come up with an alternative. But do not hold your breath. Sheikh Yamani, Saudi oil minister for nearly a quarter of a century, made the point rather memorably. "The Stone Age didn't end for lack of stone," he said, "and the oil age will end long before the world runs out of oil." The most powerful influence on oil usage in the medium term will not be the supply of oil but its price. The more expensive it gets, the more pressure this creates, even on the reluctant, to come up with viable alternatives. This is not presently the case with oil's close cousin, natural gas, which is getting cheaper, at least in some parts of the world. If transport is oil's natural metier, gas is used principally in electricity generation and heating. It is, essentially, oil-lite. Liquefied or under pressure, it has roughly 40% of oil's energy content, although about two-thirds more than brown coal. Even though oil is cleaner than coal, gas is the cleanest of the fossil fuels, producing half as much carbon dioxide as coal and two-thirds that of oil when burnt. It produces dramatically less nitrogen oxide than either and no sulphur to speak of.

There will be no immediate shortage of natural gas supplies, with big new finds in places such as Tanzania and Mozambique. Qatar, which has the world's third-largest gas reserves after Russia and Iran, keeps finding more. (The Qataris have been spending some of their new-found wealth in London where, among other trophy investments, they now own the famous Harrods department store, The Shard skyscraper, and the Grosvenor Square site of the soon-to-be former US embassy. We have dealings with their London-based head of property investment.) Then there is the emergence of fracking, which is unlocking huge quantities of 'unconventional' gas from shale rock and so-called 'tight sands' or sandstone. The technique, developed in the 1990s by Texas oilman George Mitchell, involves blasting these nonporous geological formations with water, sand, and various chemicals. It uses lots of energy and water, which its opponents do not like, and they dislike even more its potential for contaminating aquifers, releasing methane, or even causing earthquakes. Its backers say these risks can be managed, at a cost, which is worth paying since the potential is enormous. Fracking has become widespread in the US, producing large quantities of inexpensive gas and forcing down prices, to the point where US gas producers are finding it difficult to make any money. We are involved with US companies that are active in this new industry.

Cheap shale gas now represents nearly one quarter of all US gas production and, as it displaces more expensive coal in electricity generation, US carbon emissions have been falling.[11] That is ironic, since in Europe, which has tried so much harder to reduce emissions, they have risen. This is partly because European gas prices are contractually linked to the price of oil, unlike free-floating US prices, and are therefore very high. So the dirty old coal that is no longer being burned in the US is being exported to Europe, where it is displacing more expensive but not-so-dirty gas. That is global markets for you.

If the US has all the oil and gas it needs, the world could become a different place. The American Gas Association reckons that the US has nearly a century's supply of gas, half of it in shale and other rock formations.[12] More strikingly, the IEA says that, with the help of shale gas and tight oil, the US will overtake Saudi Arabia and Russia to become the world's largest global oil producer by 2017.[13] Some predict that the US could now achieve the energy independence it has so desperately wanted by 2030, or even 2020. This would have all sorts of geopolitical consequences. The US would invade the Middle East less often, for one thing. But it could also have at least one negative side-effect, easing the pressure on the security-conscious Americans to develop more non-carbon technologies.

Other countries may soon start blasting their own shale geologies, including China and Australia, both believed to have large deposits. Europe has nearly as much shale gas as America, although exploitation may prove harder for geological, legal, and political reasons. The shale gas tends to lie deeper underground, and is therefore more difficult to recover. In the US, what lies under the land usually belongs to the landowner. In Europe, it usually belongs to the state, so landowners are less keen on the sight of oil and gas rigs. Test fracking near the UK seaside resort of Blackpool produced two earth tremors in 2011, resulting in a blaze of negative publicity. The British government eventually decided to allow the practice to continue, although with additional monitoring, but France, Bulgaria, and the Czech Republic wasted no time in banning fracking for shale gas.

Oil-fired power stations can generate electricity on a large scale and can theoretically be used for baseload or load following. Given the price of oil, however, this is expensive power. So whereas a few oil-rich countries such as Saudi Arabia use oil for a significant proportion of their generation, most others prefer to keep it on standby, particularly since it can be brought on stream relatively quickly when needed.

Gas, on the other hand, has been enjoying a golden age in generation and will probably continue to do so for some time. Gas turbine plants, which can be built rapidly and fairly cheaply, fall into two categories – open circuit, and closed circuit or 'combined cycle gas turbine' (CCGT). The first is capable of very quick start-up but is relatively inefficient, and so is generally used as a standby plant. The second recycles its exhaust heat, boosting efficiencies, but takes longer to fire up. In Europe, a CCGT plant costs about $900/kW to build and construction times are typically about two years.[14] This makes them cheaper and quicker to build than coal-fired plants although, at about 30 years, they have a shorter technical life. The weaknesses of gas-fired generation include its sensitivity to changes in gas prices and, depending on where it is situated, vulnerability to any volatility in supply. That volatility can be as much political as logistical. Russia, which supplies a quarter of the EU's gas supplies, has shown itself quite prepared to turn off the taps when it feels provoked. So no one in their right mind wants to be too dependent on Russia for keeping the lights on.

NUCLEAR

Compared with electricity from coal, oil, or gas, nuclear power is 'clean', although its opponents would contest that point. But before we get into that argument, let us sketch out the basics. The process involves splitting atoms of uranium-235 by bombarding them with neutrons. The chain reaction set off by this nuclear 'fission' releases large amounts of energy that can be used to heat water, produce steam, and drive a turbine in much the same way as fossil-fuel-powered generation.

Uranium is found just about everywhere in the Earth's crust, and is present in most rocks, usually at very low levels. It has two isotopes, uranium-238 and uranium-235, the latter being by far the rarer and most radioactive of the two. Most natural uranium contains fewer than 1% uranium-235 and must be enriched to between 3% and 5% for use in power generation, or between 85% and 90% for nuclear weapons. Enrichment is generally carried out using highly sophisticated gas centrifuges. The only reason that many countries lack the ability to make nuclear weapons is building or acquiring arms-grade centrifuges is extremely difficult.

Today, the world consumes about 180 million pounds of uranium a year (excluding military consumption, which is unknown). The vast bulk of that comes from mining, and a small proportion is produced by reprocessing spent fuel rods. Some 13%[15] comes from the US-Russian Highly Enriched Uranium Agreement – otherwise known as the Megatons to Megawatts Program – that turns old Soviet nuclear warhead uranium into low-enriched uranium for power plants. The largest uranium mine in the world is at McArthur River in northern

Saskatchewan, Canada. McArthur River has abnormally high-grade deposits, containing up to 20% uranium-235. That is sufficiently radioactive for the mining to be carried out by remote-controlled robots. Australia has the world's largest-known uranium deposits but Kazakhstan is the biggest mining producer, having overtaken Canada in 2010. Australia comes third in the production league, followed by Niger, Namibia, and Russia. There will be no shortage of uranium any time soon.

Radioactivity is, of course, uranium-235's dark side. The prospect of radioactive fallout as a result of nuclear accident or catastrophe – as at Chernobyl or Fukushima, for example – is cause for public concern, although, as we shall see, it frightens people more than is really justified. Even so, exposure to radiation can cause increases in cancers and, after very high doses, radiation sickness and death. Then there is the problem of how to dispose of spent but still radioactive nuclear fuel safely.

Until now, there has been no practical substitute for uranium-235. In the future, however, a safer alternative may be thorium, which has certain very desirable advantages. It is even more abundant than uranium, does not require enrichment, and is much more difficult to use in weapons. Unlike present uranium reactors, which are giant pressure cookers, the molten salt reactor designs proposed for thorium operate at normal pressures. They can use nuclear waste from other reactors as fuel, helping to solve, rather than adding to, the waste problem. Perhaps most importantly, the fission process is much easier to shut down and most of what little waste they leave is safe within a decade. Various countries have been experimenting with thorium-based reactors, including India, which has extensive thorium deposits, and China. Thorium research still has some way to go, although some believe we will see working thorium reactors by 2028. We shall be taking a closer look at thorium later.

Nuclear plants are well-suited to baseload generation. They take a while to start up, but they can have very high load factors and are very cheap to run. Unlike oil and gas, where large and unexpected price rises can suddenly make a project uneconomic, ongoing fuel costs for nuclear plants are a small percentage of the whole. So even if the uranium price doubled, the economic effect would be relatively slight. The overwhelming cost burden of nuclear projects lies in the up-front capital expenditure – they are slow and very expensive to build. The IEA puts typical nuclear build costs in Europe at about $4,000/kW[16], and typical build times at five years. The financial estimates will have been inflated by the experience at Finland's Olkiluoto 3 project. Western Europe's first new nuclear power station for 15 years, this has run disastrously over time and budget and looks like taking 10 years to build. Most nuclear project budgets also have to take decommissioning costs into account, which can average $800/kW, and waste disposal. In Finland, however, the state assumes liability for nuclear waste, funded by a small levy on the price of nuclear electricity. Nuclear's running costs are among the cheapest of all and, if and when added carbon costs begin to kick in for fossil fuels, the all-in cost playing field will be levelled more in nuclear's favour.

As things stand now, the costs and risks associated with nuclear mean that new private-sector-funded plants are unlikely to be built without some form of public help. And even though public subsidies are permitted for 'renewable' energy projects within the European Union, nuclear is not classified as renewable. Opposition to nuclear has recently hardened among some member states, notably Germany, making the possibility of reclassification more remote. Although Olkiluoto has not done nuclear's reputation any favours, it was the meltdown at Fukushima that delivered the latest threat to its prospects. The earthquake and tsunami that devastated the north east coast of Japan's Honshu island in March 2011 knocked out the nuclear power plant's cooling pumps, causing a number of reactors to overheat. The television pictures were harrowing, but the situation was eventually brought under control and there was no loss of life. Nonetheless, the disaster prompted reaction on an international scale. The most extreme knee-jerk response came from Germany, which immediately closed down all pre-1980 reactors and ordered utilities to shut down all the others by 2022. Japan shut down all its own reactors and is reviewing its nuclear policy. Switzerland and Italy have imposed nuclear moratoria and various other countries, such as Brazil, have delayed their nuclear plans. This is the kind of political risk that makes nuclear, without any guarantees or sweeteners, even less enticing to the private sector. China, on the other hand, is unsentimentally accelerating its development of nuclear power, alongside shale gas and – our next subject – hydroelectric energy.

HYDRO

We said that there was no perfect energy option, but hydroelectric energy comes close. The trouble is that only certain countries are lucky enough to have the topography and the water supply to make it happen. Although it is utterly 'renewable' and as 'clean' as it gets, hydro is included in this chapter rather than the next because it is such a tried and tested part of our existing energy mix.

Like fossil-fuel generation, hydro converts one form of energy into another by turning a turbine-mounted magnet inside a coil. The difference is that it uses the kinetic energy of falling or flowing water rather than the thermal energy of heat and steam. Although it is possible to generate on a small scale with 'run of river' plants, relying on river flow only with no reservoir, most hydro comes from dammed water. The amount of power that can be extracted depends on the 'head', the difference in height between the upstream water level and the outlet level – the higher the dam, the bigger the power capacity.

Unlike other renewables such as wind and solar, hydro can deliver power on a huge scale. And unlike them it is predictable and controllable and can be switched on and off whenever it is needed. This almost instantaneous availability makes it highly suited to standby generation or load following. Countries with abundant water resources are more likely to use it for baseload – such as Norway, which gets 99% of its electricity from hydro, Canada (57%), Switzerland (55%), and Sweden (44%). Another positive feature of hydro is the possibility of 'pumped storage', which can boost revenues for generators and help with load management. On the same principle as a storage heater, this uses low-priced electricity at times of low demand – late at night, perhaps – to pump water back up into smaller reservoirs. From there, the water can be released at peak demand periods to generate more power for sale when prices are highest.

Hydro is not entirely perfect, however, and it does have its downsides. Large schemes can displace thousands of people, who have to pack up and move as rising waters in the new dam cover their homes. They can cause problems for farmers downstream, by interfering with their water supply, and play havoc with fish populations. Downstream water quality may be affected and, if the dam fails, the results can be catastrophic. Hydro projects can take up to eight years to build and are very expensive, although the costs per kilowatt fall rapidly as the capacity rises, from $3,900/kW for a small European scheme to $2,230/kW for a large one.[17] Depending on their situation, the electricity may have to be transmitted over long distances, adding to costs. Hydro installations can last 100 years, and the fuel is free, but they are dependent on the availability of water, and some warn that climate change could affect future hydro generation.

Even with those caveats, hydroelectricity does just about everything that environmentally-minded legislators want from power generation and it can do it on a large scale. Unfortunately, since it requires sufficient water supply and the right terrain in the right (ie underpopulated) place, the technology is not reproducible at will. For countries not blessed with those virtues, other clean – if less mature – technologies seem more accessible and we shall now examine some of their strengths and weaknesses.

In the following chapters, we shall look at the strengths and weaknesses of alternative energy sources, and review developments in the world of transport. We shall meet some of the people who are creating green energy projects and then we'll examine how Bubbles start. Because energy saving is the most effective form of green energy, I have included a chapter on energy efficiency. Then I explain how European energy policy is effectively a very large bet on rising fossil-fuel prices, a bet that we may be destined to lose. I show how the green energy Bubble is already leaking air before concluding with my own prescription for what we need to do to keep the lights on.

CHAPTER 2:
A CLEANER HABIT?

Our capacity for self-reflection is one of the things – perhaps the only thing – that distinguishes us from other living creatures. And the more we know about ourselves and our surroundings, the more we are inclined to worry. We started worrying about the effect of (non-martial) human behaviour on our planet in the 1970s when we discovered an expanding hole in the ozone layer that protects us from the sun's most damaging rays. The problem was traced back to the widespread use of chlorofluorocarbons (CFCs) in aerosol sprays and refrigerants.

An unprecedented international agreement to phase out CFCs, expressed in the first United Nations treaty ever to gain universal ratification, has begun to reverse the process of ozone depletion, and the experience has left people with two lasting impressions. One is that our planet and its atmosphere are a precious but delicate resource which we could destroy through careless stewardship. The other is that, if we all pull together, we can do what is necessary to sustain and protect it. Anxieties about greenhouse gases and climate change have arisen against this backdrop.

The dangers of losing the ozone layer were undeniable, scary, and clearly understood. Heightened exposure to high-frequency ultraviolet rays increases cancer and cataracts in humans and causes reproductive problems in other life forms. Climate change is less clear-cut. On the face of it, it is the ozone problem in reverse. The idea is that accumulating greenhouse gases – such as carbon dioxide and, ironically, ozone itself – create an insulating atmospheric barrier that reflects heat back onto the earth instead of allowing it to escape into space, as it should. The result, we are told, is a gradual rise in world temperatures, with an overall warming of climate and everything which that entails, like more widespread drought and rising sea levels caused by melting icebergs.

If the planet is heating up, there is no consensus on what or how serious the long-term effects might be, or to what extent humans might be responsible. But the disagreement does not start there, because not everyone accepts that the climate *is* indeed changing over the long term or that we are to blame. Theories range from the imminent doom preached by eco-warriors to global warming as a communist plot. The latter is the contention of James Delingpole, an articulate right-of-centre English journalist, in his book *Watermelons* (green on the outside, red on the inside), which is not quite as wacky as it sounds.[18] This, however, is not the place to argue the toss about climate change. Let us simply acknowledge that government actions on anthropogenic or man-made greenhouse gases in general, and carbon dioxide emissions in particular, for reasons of genuine concern as well as electoral appeal, are facts on the ground. Whatever the status of the climate change debate, this politically inspired behaviour is already having a determining effect on our energy future.

Beyond a general acceptance that man-made greenhouse gas emissions are not really a good thing, the international governmental response has been markedly less unanimous than it was in the case of CFCs. The challenge and its implications for policy and the national purse are much more daunting and complex than simply coming up with a replacement for hairspray propellant, and the lack of scientific consensus has allowed some wriggle room for the more unwilling. Even so, many governments have set targets for carbon emission reductions and, in some cases, increased the use of biofuels.

In March 2007, the European Union (EU) adopted its 20-20-20 targets, setting three objectives to be achieved by the year 2020. They were a 20% reduction in EU greenhouse gas emissions from 1990 levels; a 20% improvement in the EU's energy efficiency; and raising the share of renewables in EU energy consumption to 20%. Individual member states are free to set higher targets if they wish. In the case of renewables, each state has its own individual target, ranging from 10% in Malta to 49% in Sweden. 'Renewables' include hydro, biomass, wind, and solar power but, crucially, not nuclear power.

In 2012, renewables contributed 12.4% of the EU's energy consumption[19], up from 103% in 2008. Sweden led the member states with 51%, beyond its 2020 target. Denmark was running at 26% (2020 target is 30%), and the UK was close to the bottom of the list at 4.2% (2020 target is 15%). Norway, although not an EU member, was top of the class at 64.5% (2020 target is 67.5%), thanks to its abundance of hydroelectric power.

One integral, although so far not very successful, part of the 20-20-20 package is the EU's Emissions Trading System (ETS). This sets an upper limit on the greenhouse gases that can be emitted by the European power plants and factories included in the system. The limit, or cap, reduces over time. Alongside this shrinking cap is a system of 'allowances' or 'credits'. Each year, the companies concerned must obtain enough credits to match all of their emissions. They are allocated credits equivalent to the level of their official cap. Like most things connected with the EU, it is a complex system but, whereas the credits were allocated for free to begin with, over time more of them have to be paid for. Next comes the interesting bit. Companies emitting more than their cap allows must now buy extra credits in the market. Those emitting fewer than their targets can sell their leftover credits – which is why it is called a 'cap and trade' system. The theory is that the pain of having to pay for credits will goad transgressors into reducing emissions. Electricity generators, for example, should be persuaded to invest in more renewable generation.

The carbon credits have a price, which moves up and down according to supply and demand as they are traded on various exchanges around the world. ETS is the largest emissions trading system, but not the only one. Australia, New Zealand, and the US state of California each have their own and the US operates cap and trade markets for other pollutants. Korea is setting up a carbon system and the hope is that, eventually, different carbon credits will be recognized and traded internationally. When the EU scheme began, however, its administrators made a very silly mistake. Having been lobbied to distraction by industry, German steelmakers not least among them, they gave away far too many free allowances. This abundance of supply coupled with the economic downturn has driven initial prices from nearly €30 per tonne of carbon emitted, down to a brief but record low of €2.81 in early 2013. With carbon prices at such low levels, big power utilities have found it cheaper and more convenient to buy the carbon credits they need each year in the market, rather than building their own green energy capacity or fitting carbon capture and storage to their fossil-fuel plants. Indeed, shortly after prices fell below €3, the *Wall Street Journal* reported that utilities in the Czech Republic, Germany, and Poland were "reconsidering plans to phase out coal-fired plants".[20]

China is the biggest emitter by some margin, with 28% of the world's carbon emissions in 2013[21], more than the US and EU combined (US 14%, EU10 %, and India 7%). It has gradually come round to accepting the need to rein in emissions, thanks to a combination of worsening air quality, foreign pressure, and growing sensitivity to its dependence on fossil-fuel imports, such as coal from Australia and oil from the Middle East. China's 12th *Five-Year Plan*, published in 2011, promised to raise nonfossil-fuel energy from 8% to 11.4% of the total by 2015. It also aims to install 160GW of wind and 50GW of solar power by 2020. In the US, where targets are known as 'renewable portfolio standards', the picture is more jumbled, since each state has its own standard. Connecticut aims to have 27% renewables by 2020, for example, whereas Arizona is going for 15% by 2025.[22]

Setting targets is one thing, but achieving them is quite another. You can wave the stick at energy producers but, to persuade existing generators to embrace clean technologies and new ones to enter the field, it helps to offer a carrot as well. Big utilities can be clubbed into compliance, but without the carrots there would have been no new entrants to the solar and wind industries, simply because these were not – and, for the most part, are still not – economic propositions in their own right. Some countries, including the UK, the US, and Australia, use 'renewable energy certificates' (RECs) or 'renewables obligation certificates' (ROCs) to motivate and enforce, like carbon credits in reverse. The technicalities vary but the principle is to award generators with certificates for every unit of renewable energy they produce. Each year, the big utilities must collect enough certificates to show that they have met clean energy targets. Although they can acquire them by generating green energy themselves, in practice they have largely preferred to buy them from new solar and wind producers.

Although critics say that this process of offsetting does nothing to curb existing emissions, the certificates do promote new green capacity by acting as an additional source of revenue for clean generators. They are also an added cost for polluters, making cheap fossil-fuel generation more expensive for them and, therefore, making alternatives more economically attractive. In the US, some states have used the multiplier principle to promote a particularly favoured technology, awarding three certificates per MWh of wind energy, say, compared with one for solar. Others have solar 'carve-outs' or 'set-asides' that set a specific standard for solar in addition to the umbrella requirement for renewables in general. In Scotland, wave and tidal generation gets five certificates for every one earned by most other renewables.

If we want cleaner fuels, they must either pay their own way as competitive businesses or we must subsidize them. At their current levels of technical performance and cost, few green sources can compete with fossil-based fuels without financial help. RECs are a great idea, but on their own they are not enough to attract private capital, and so other forms of subsidy have been given to green projects. These include government-guaranteed (and therefore cheaper) loans; tax breaks; and, most commonly, paying higher than normal prices or 'feed-in tariffs' for green electricity. Whatever form the subsidy takes, it is the taxpayer or consumer – you and me – who ultimately pick up the bill. This may be through higher taxes; the opportunity cost of our taxes being diverted from other, perhaps more desirable uses; or through higher electricity prices.

When it comes to raising money for green projects, many of them find it difficult to attract the attention of the large investment banks. The Morgan Stanleys and Deutsche Banks of this world only deal with the big boys, and there are only so many wind farms that a major utility such as Denmark's Dong Energy can build each year. And in my experience, clients from the alternative energy sector are very different from, say, oil and gas industry players in one respect, which is that they are very stingy when it comes to paying fees. They are so used to getting everything for free that they seem to expect free banking as well. That said, their sums are so finely balanced that, if we charge too much, we can bring down the whole economics of a project. Now, let us examine the relative attractions of these alternative technologies in more detail.

SOLAR

The sun is the most powerful energy source we have by far. It would provide as much energy as we could ever want if only we knew how to harness it. The energy that hits the earth in only 15 minutes of solar radiation would satisfy the whole world's needs for a year. Indeed, most of the energy sources available to us are essentially 'solar'. Fossil fuels are solar energy in long-term storage. It is the sun that drives wind and water cycles, and provides the energy for plants to grow, so hydro, wind, wave, and biomass are all derivative solar energies of one kind or another. Only geothermal, which relies on the earth's inner heat, and tidal energy, driven by the moon, are not.

When we talk about 'solar energy', however, we usually mean using direct sunlight to generate heat or electricity. Sunlight falls everywhere and can provide energy even on cloudy days, but its intensity, and therefore the amount of energy it provides, varies according to cloud cover, latitude, and time of year. Obviously, the Sahara Desert gets more sunlight than the Arctic. But relatively dry southern Spain may get more sunlight than the wetter, and therefore cloudier, parts of Central Africa. Wherever it falls, the sunlight is, of course, free of charge.

Direct solar heating is at the simplest end of the technology spectrum. Here, sunlight heats a circulating fluid for the purpose of heating an enclosed area. This works well for low-temperature applications, and home and business installations have grown rapidly in recent years, with installed capacity already past the 100GW mark.[23] Growth is particularly noticeable in China. Easy to install and relatively cheap, 'solar thermal' reduces the need for other power sources and has been described as the "unsung hero of green energy". But it is unlikely to make a large impact in closing the energy gap.

A more complex and potent variation on the solar thermal theme is 'concentrating solar power' (CSP), which uses mirrors to focus sunlight on a single concentrated area. The result can heat liquids – often molten salts – to much higher temperatures, in some cases up to 1,000 degrees Celsius. This will produce steam to drive a turbine, or heat gas to drive a Stirling piston engine. The resulting rotary motion is used in time-honoured fashion to turn a magnet and so generate electricity. The most mature CSP technology designs the mirrors in long parabolic troughs. Every bit of the mirror's surface reflects the sunlight onto a pipe that runs down the middle, heating the fluid that it contains. Higher temperatures can be achieved if the mirrors are flat, with arrays of them mounted on sun trackers in towers – 'solar tower' technology. The temperature is higher still with parabolic dishes, which have a Stirling engine installed at the focus of the dish.

CSP systems need high levels of direct solar energy and can only be built in sunbelt regions such as southern Europe, North Africa, or the south-western US. The good news is that they can generate electricity on a utility scale that can make a meaningful contribution to the grid, and the IEA reckons that CSP could provide up to 11% of global electricity by 2050.[24] Large-scale CSP tower and parabolic trough plants have been built in the Mojave Desert and Andalucia, among other places, and the World Bank is supporting projects in Algeria, Egypt, Morocco, Tunisia, and Jordan. With an investment cost of $4,000 to $8,500/kW, depending on the specific technology, CSP still needs subsidy to compete, but costs should reduce substantially as the industrial production volumes of CSP components rise. Load factors range up to approximately 28% at best.

The third and, to date, fastest-growing form of solar energy is photovoltaics (PV). PV relies on the photovoltaic effect first observed by Alexandre-Edmond Becquerel in 1839. He discovered that, under certain circumstances, some materials become 'semiconductors' and produce electricity when exposed to light. Silicon is the most widely used semiconductor in PV solar technology, assembled in cells to generate electricity that may be connected to the grid or off-grid. Today, the most widely used semiconductors are silicon and gallium arsenide.

There are two basic ways of making silicon panels: crystalline silicon and thin film. With the first, silicon crystals are grown to a high degree of purity and then sliced into wafers. Crystalline cells, which account for about 80% of the PV market, are more efficient (converting up to 20% of the solar energy into electricity) but also more expensive than thin film, and take up less space. Module costs in Europe range up to $2,850/kW.[25] Thin film silicon is exactly that – a thin film of silicon atoms. At up to \$2,000/kW, it is cheaper to make than the crystalline product but has relatively low efficiency (less than 10%) and so requires a larger area for every watt of electricity produced.

Solar PV has recently been the fastest-growing of the renewables, with a compound annual growth rate of 56% over the past five years and a doubling of installed capacity between 2009 and 2011, according to market intelligence firm GlobalData.[26] Europe, where it has been heavily subsidized via feed-in tariffs, has been the largest market. However, European subsidies are on the decline and the momentum has shifted to emerging markets such as China, Japan, and India, which have announced ambitious PV targets. India is short of energy – hence the blackouts – and I have recently had discussions with several very wealthy Indian business families who are interested in investing in the green space. In Europe, Bulgaria was a hotspot for solar investment but is no longer, and has been replaced by Romania. The subsidies now available in the UK have made it increasingly attractive to entrepreneurs – and it is probably safer than Romania. Everywhere, an important growth driver has been a fall in equipment costs as manufacturing has migrated eastwards, mainly to China. That has precipitated the collapse of a number of US and European solar PV manufacturing businesses which are no longer able to compete. We shall hear more about this later.

In the electricity industry they talk about 'grid parity', which is the point at which the cost of producing an alternative energy is the same or better than the normal cost of buying electricity from the grid. From then on, subsidy should no longer be required to make the technology competitive, so grid parity is a destination that governments would very much like the solar industry – and all other subsidized green energy – to reach. GlobalData expects some US PV projects to have reached grid parity by 2014 and most others in the US by 2017, helped by overproduction – and hence falling prices – of PV modules. China, it says, will hit grid parity in most regions by 2015–2016. Intermittency remains solar's big problem, however, with PV load factors of up to 22% or 23% only in the sunniest locations.

GEOTHERMAL

At first glance, geothermal power represents an enormous, largely untapped resource. We are talking about the heat energy that lies beneath the earth's surface, originating in its molten inner core. Until now, however, the heat has only come close enough to the surface to be useful in areas where the continental plates are colliding. The Pacific Ring of Fire is an obvious example, stretching up the western seaboard of the Americas and down through Japan, Indonesia, and New Zealand. Iceland is another. This tectonic activity gives us earthquakes and volcanoes but, more helpfully, it also heats up underground reservoirs of water that can be used to provide heat or, at higher temperatures, power.

'Dry steam' plants feed directly off underground steam, piped up and used to power a turbine, as at The Geysers, north of San Francisco in the US, where they have been producing clean, reliable energy on a commercial basis for more than 50 years. Well-established dry steam plants can also be found in Iceland and parts of Nevada. More widely used today are 'flash steam' plants, which tap into underground water, under pressure at temperatures up to 180 degrees Celsius. When brought to the surface, pumped up or flowing under its own pressure, it 'flash' boils, producing steam that can be used to drive electricity turbines or piped off to heat nearby buildings. Geothermal power is cheap and provides baseload supply, in that it can operate 24 hours a day all year round, or at least until the subterranean pressure starts to fall. And yet the installed base is only about 11GW worldwide, limited as it is by location and the fact that individual plants tend to be rather small.

One emerging technology promises to let us harness geothermal energy at locations that were previously thought impossible. It is called 'enhanced geothermal systems' (EGS) or, more colloquially, 'hot dry rocks'. As the name suggests, hot rocks is available just about anywhere, depending on how deep you are prepared to drill for it. A well is drilled down to the hot rocks, which will be some way below the water table at depths of 3km or more. Pressurized water is injected into the well to create or widen fractures within the dry rock, to create a reservoir through which water can pass – this is a form of 'fracking' and the most controversial part of the process. Then a second, 'production' well is drilled into the fractured area, so that water can be circulated, heating up along the way and used to generate energy back up on the surface.

The potential of EGS is immense and it caused much excitement among investors in 2009, when shrewd project managers began to market it as 'green baseload'. For a while, EGS ventures raised considerable amounts of money. The technology does, however, have one slight drawback. It can cause earthquakes. In fact, the fracking process routinely does cause micro earthquakes. The trouble starts if and when the pressurized water interacts with existing deep faults and kicks off a major seismic event. This doesn't happen very often, but when it does it does little to advance the technology's cause. EGS is more profitable if it generates heat as well as electricity, in which case the customers have to be nearby so that the heat can be piped to them without too much loss. But if EGS is proposed near people's houses or workplaces, in an urban environment, the seismic issues become a hurdle. The dangers became apparent when an EGS project set off an earthquake (3.4 on the Richter scale) in Basel in 2006. Basel, surprisingly enough, has a history of earthquakes – in 1396 the city was rocked by the largest earthquake ever recorded in central Europe (Richter 6.7) and badly wrecked. The 2006 event did almost as much harm to EGS's reputation, even though the damage to Basel itself was relatively slight, and the Swiss project has been on permanent hold ever since.

Scientists believe that these risks can be managed, at a cost, but it will be some time before geothermal can make a significant contribution to energy supply. The IEA estimates the capital costs of European geothermal at $2,400/kW, although lower where geological conditions are more favourable, such as Iceland.[27] There are operational EGS plants at Soultz-sous-Forêts in Alsace and, sitting on similar geology 40km away over the German border, at Landau, but both produce power on a very small scale. The US Department of Energy believes that the US could produce 100GW of geothermal electricity within 50 years with the help of EGS.[28] That may be so, but geothermal is highly risky, since drilling can take several years and the power output is far from guaranteed until the project is complete. So you could plough many millions into a project and only discover at the end that you will not be able to sell enough electricity to make it pay.

Australia has near-ideal conditions for EGS and has been handing out a lot of free money for renewables projects. One of the beneficiaries, and at one time the darling of the Australian EGS business, is a company called Geodynamics. Technical problems, a well blow-out, delays, and cost overruns have caused a key partner in most of Geodynamics's ventures, Origin Energy, to write off all of its EGS investments. Geodynamics' share price has fallen from more than A$2.00 in 2008 to about four cents. They were supposed to be the best in the business so, if even they are having these problems, the business is clearly a tricky one. It is this kind of risk that holds back the development of various alternative energies. On paper, geothermal is a cheap, clean, and promising energy source but it will not be riding to our rescue for some time yet.

WIND

A platoon of wind turbines braced across an otherwise clear horizon may symbolize the future of energy for some, even as it infuriates others. But the sight of windmills on the skyline was no novelty for our ancestors. We have been harnessing wind energy for millennia, first to power sailing boats and later to grind grain and pump water, long before we ever discovered coal and oil. The stumps of old windmills dotted across Holland, Denmark, and the UK should remind us of that.

Wind is yet another product of solar radiation striking the earth. The sun's rays heat different parts of the planet to different temperatures. The warmer the air, the higher the air pressure and, since air moves naturally from high- to low-pressure areas, it creates wind, which has useful energy – clean and, at least as far as the fuel is concerned, free. Winds are strongest in areas that allow unimpeded airflow, which means that the open seas are the best place to erect wind turbines – or they would be if it was practical to transmit the electricity over such long distances to where it was needed. For the same reason, onshore wind speeds increase as you get higher above the ground surface. So tall wind towers generate more energy than short ones, whose wind is slowed by the terrain.

The first windmill to produce electricity was built in Scotland in 1887, using cloth sails, followed shortly by a much larger machine in Cleveland, Ohio, with 144 metal blades. Danish scientist Poul la Cour built his first experimental wind generator in 1891 and went on to earn an international reputation as a windmill inventor. As a country with a lot of wind, Denmark took to the technology enthusiastically.

The size of the blades and the wind speed are the two key factors determining how much energy a wind turbine can produce. The longer the rotor blades, the larger the area they sweep. Since more air passes through them, more energy is produced. Longer blades require taller towers that, as we have said, benefit from higher wind speeds. Economies of scale work in the developer's favour, since the increase in output from bigger kit generally outweighs the increase in cost. That explains why wind turbines have been getting larger. Thirty years ago, they typically had 15-metre diameter blades and 50kW output. Today, 120-metre blades – as high as a 40-storey building – and capacities of 6,000kW (6MW) are not uncommon.

ca 1980   ca 2010

It should go without saying that wind turbines need to be erected in windy places. Prevailing wind patterns mean that the best sites are in western coastal regions in the northern hemisphere, and near eastern shores in the southern hemisphere. You need wind speeds of 10mph or more to make power generation practical and, because there is a cubic relationship between wind speed and energy, a doubling of the wind speed gives you eight times as much energy. So windier is better, but only up to a certain point. Wind turbines reach maximum electricity output at about 33mph and when the wind blows faster than 50mph they must be shut down to avoid damage to the rotor mechanics. Because energy is a cube of wind velocity, and because velocities in a single location can vary significantly over short periods, the output from a single turbine can be highly erratic. On a wind farm with many turbines, wind speeds can differ from tower to tower, which tends to have a smoothing effect on total output.

A flurry of onshore wind construction has been supplanted by more focus on offshore installation, to take advantage of the windier conditions at sea. Offshore towers can be considerably bigger than land-based units and a new generation of floating turbines can be installed in deeper waters, out of sight of the land and away from shipping lanes and fishing grounds.

The wind may be free, but wind farms cost a lot of money to build, especially if they are out at sea where they will generate more power. Typical construction costs per kilowatt in Europe are about $1,700 onshore and $3,400 offshore.[29] With the stresses caused by the wind, maintenance costs are relatively high, and considerably higher still in offshore installations. One of wind energy's problems is that windy places are often remote places, so that transmission lines must be built and paid for if the power is to be sold to the grid. Nonetheless, wind has been targeted as a key renewable, particularly in Europe, where it has been encouraged by high feed-in tariffs that utilities are obliged to pay producers. Renewable energy certificates offset some of the costs. Germany, Spain, and Denmark are among the leading European users of wind energy, and the UK has as much offshore wind capacity as the rest of Europe combined. In the US, oilman T Boone Pickens insists that the country could generate 20% of its electricity from wind.

We shall take a closer look at some of wind energy's negatives in a later chapter, but the chief one is that the wind does not blow all the time, and it does not necessarily blow when electricity is needed most. The average offshore load factor in the UK in 2013, for example, was 29%.[30] Although many people hate the sight of them, they also take up a lot of space. Hewitt Crane calculates that to produce one cubic mile of oil a year from wind, you would have to cover one third of the land area of the European Union with wind turbines, at a cost of some $6 trillion. The T Boone Pickens scheme would occupy a swathe of land up to 20 miles wide stretching 2,000 miles from Texas to the Canadian border.

WAVE

Wind also gives us waves, transferring to the seas and oceans the energy it gets from the sun. Waves can travel long distances, from mid-ocean to seashore, without losing much of that energy and the harder the wind blows, the higher and faster the wave. A wave's energy is proportional to the square of its height, so a two-metre wave has four times as much energy as a one-metre wave.

In terms of energy potential, the oceans are a powerhouse. According to some calculations, waves heading east across the Atlantic contain up to 40kW per metre of exposed coastline. The question is how to extract it. Viewed in cross-section, wave action is a continual raising and lowering of the water's surface and wave-energy projects seek to harness this movement, in ways ranging from the imaginative to the bizarre. None of these technologies could be described as mature – which is why there are so many of them – and all are hugely uneconomic right now. That has not stopped some countries setting targets for wave and tidal energy supply. In the UK, surrounded as it is by water, the government wants 300MW of marine energy capacity by 2020 and the UK industry says that the seas could one day generate 20% of the country's needs.[31] Portugal, with its Atlantic coastline, has targeted 250MW of installed wave capacity by 2020.[32] It also established what it called 'the world's first wave farm' at Aguçadoura, north of Porto, in 2008. Sadly, the 2.25MW facility was shut down after two months, due to technical problems.

Marine energy has attracted the attention of scores of developers – about 80 companies worldwide, according to Bloomberg New Energy Finance. Many are developing wave systems based on buoys, whose bobbing or pivoting movement is then transformed into energy, and many claim these can be scaled up to utility-sized output. The equipment used at Aguçadoura was the Pelamis, perhaps the best known and certainly one of the more exotic of the emerging technologies. In the natural world, 'pelamis platura' is the Latin name for the yellow-bellied sea snake, which gives you an idea of what the machinery looks like. It is a semi-submersible, articulated or hinged metal snake that faces into the waves and undulates with the surface of the water. The joints, which flex up and down, contain hydraulic power take-off systems that convert the movement to electricity. It was unexpectedly heavy wear in these joints that brought the Aguçadoura project to a halt – Pelamis has since been working on a second-generation snake. Remote control from the shore, or even from a distant location, allows generation to be maximized or minimized, depending on the weather, and the snake ducks too-big waves "by diving through them, just like a surfer heading out to sea", according to its designers.

Ocean Power Technologies of Pennington, New Jersey, is one of the leading wave powerhouses. Its PowerBuoy – which looks just like a buoy – moves up and down with wave action and converts the kinetic energy to drive an electrical generator. OPT has had a number of pilot projects, past and present, off the coasts of Hawaii, Oregon, Spain, and Scotland, among others. Scotland is a hothouse of marine energy research, energetically supported by the Scottish regional government. It is the home of Pelamis Wave Power, makers of the snake. Inverness is the location of AWS Ocean Energy, which has designed a doughnut-shaped floating ring made up of air cells. Wave movement compresses the air to produce pneumatic power, which is converted to electricity. Edinburgh-based Aquamarine Power is developing the Oyster, a hinged flap on the seabed in near-shore waters, as it moves back and forth, it pumps high-pressure water ashore where it drives a hydroelectric turbine. Then there is the more substantial Limpet, a concrete chamber built into a coastal rock face – the first has been built on the Scottish island of Islay. Incoming waves push water up the chamber, forcing air into a low-pressure air turbine, known as a Wells turbine.

A number of these technologies have been deployed in pilot projects off Scottish islands such as Islay and Orkney. Australia's Carnegie Wave Energy has a joint venture with French utility EDF for a project off the coast of Reunion island. At this stage of development, wave energy is probably best suited on a standalone basis for island communities, where building other forms of power plant is impractical or expensive. In the long term, it could conceivably be hooked up to the grid.

Although load factors can range from 60% to as little as 25%, lower even than some wind installations, wave activity can be predicted some days in advance, which would allow the grid to plan ahead. Visual impact, so contentious with onshore wind farms, is low or non-existent. But it is very expensive both to construct and to maintain, involving tons of machinery in a hostile and corrosive environment. The investment cost can range from $6,800 to $9,000 per kilowatt[33] and there is much scope for things to go wrong, as the Aguçadoura project demonstrated. Denmark's Wave Star project ended in bankruptcy, despite being funded by one of that country's more successful businessmen. Trident Energy of the UK spent years of work and millions of pounds on a prototype wave generator only to watch it sink as it was being towed to its trial site. That is one of the reasons why private sector investors are wary of wave projects, which tend to rely heavily on support from public funds.

TIDAL

If the seas are a conduit for solar energy, they also bring us energy from our nearer neighbour, the moon. Twice a day, lunar gravity hauls the tides in and out, producing powerful incoming and outgoing lateral movement, as well as up and down movement as the sea level rises and falls. Those movements can be more powerful still if the tide is running through a narrow channel, such as a strait or inlet.

These two distinct characteristics – horizontal tidal 'stream' and vertical tidal 'range' – allow energy to be harvested in two different ways. Tidal stream is still in its infancy. The devices come in many shapes and sizes but most involve some kind of turbine whose blades are turned by the lateral water flow. Although tidal currents move relatively slowly, water has many times the density of air, and a current of only 2 metres per second has the same energy potential as a windspeed of 18 metres per second. Some tidal stream designs use a rotating Archimedes screw. Then there are oscillating hydrofoils that flap to and fro in the current, producing hydraulic power. Strangest of all is a kite-inspired design being developed by Gothenburg-based Minesto. Invented by a former Saab turbine blade engineer, the Deep Green generator is mounted on a hydrodynamic wing and tethered to the seabed, allowing it to 'fly' around like an underwater kite in circles or figures of eight. Because it is moving, the water flow is accelerated, so it can operate in areas where the current is too sluggish for other designs. Tidal stream investment costs range from $6,000 to $7,800 per kilowatt.[34]

To exploit tidal range you build a barrage across an estuary or a bay, much like a hydroelectric dam. The incoming tide flows through the sluice gates, filling the inner basin until high tide, when the gates are closed. As the tide withdraws, the height between the sea level and the stored water – the 'head' – increases. Then the turbine gates are opened, and the stored water flows out, turning the (often very large) turbines and producing electricity. There are variations on this theme, but that is the general principle. The up-front cost of tidal barrage is not too far removed from mid-sized hydro, at about $5,000 to $5,500 per kilowatt.[18]

A few barrage projects already exist or are being planned. A 24-turbine, 240MW tidal power station has been operating on an estuary of the Rance river near St. Malo in Brittany since 1966, with a load factor of about 40%. People have been proposing a barrage across the mouth of the River Severn between England and Wales since the mid-19th century, although not, at first, for the purpose of generating electricity. The original idea was to create a harbour and provide protection against floods. Power generation was first mooted in 1925 and rejected on economic grounds. Today, the objections are more environmental – a major negative for barrages is the damage they do to wildlife and fish populations. They also have a habit of silting up. Nonetheless, the Severn estuary has the second-highest tidal range in the world and could generate up to 5% of the UK's electricity, the government says. To do so, it would have to get past a barrage of fierce environmentalist opposition. There is, however, a project afoot to build an artificial tidal lagoon in Swansea Bay, at the mouth of the Severn Estuary. With up to

26 bi-directional turbines, it would have a capacity of 220MW, taking advantage of the estuary's range while not disturbing the river itself.

The highest range is in Nova Scotia's Bay of Fundy, where a 20MW-capacity plant has been running since 1984. There is slightly more interest in the technology in the present climate. However, the location has to be right, which narrows down the possibilities rather dramatically. And its attractions are further dimmed by the fact that, although the cost is comparable to hydroelectric dam projects, the load factor is half or less, because it cannot run continuously. The days of economic generation are probably further away for tidal energy than for any other renewable technology.

BIOMASS

Biomass is staging a comeback. As the subset of energy sources that contains all living or organic matter, biomass includes wood, the main source of heat and, up to a point, light for most of our history. As the Food and Agriculture Organisation of the UN points out, wood is still the most important renewable, providing more than 9% of the world's primary energy.[36] That is as much as all the other renewables combined, including hydro. Municipal and industrial waste, together with agricultural crops and by-products, add another 1%, giving biomass overall a 10% share of primary energy.[37]

For a large part of the world's population, wood has never been displaced and remains indispensable for warmth and cooking. And even for the rest of us it is never far away. The oil crisis of 1973, when the price of oil quadrupled, prompted a surge in US sales of wood-burning stoves, and the publication of a new, if short-lived, magazine – *Wood Burning Quarterly*. High fossil-fuel prices have rekindled interest in wood more recently, this time in the shape of wood pellets. Made from compressed sawdust or other wood-processing waste, their density and low moisture content gives them a high combustion efficiency and their cost per energy content is lower than any fossil fuel except coal.

Cost and eco-friendliness have made wood pellets a popular domestic fuel for heating boilers and stoves (pellets made from grass and other biomass can be used for the same purpose). Thanks to its tax on fossil fuels, Sweden consumes more pellets than any other European country, although they are catching on in places such as Austria and Denmark. What is really beginning to drive the pellet industry, however, is their use as fuel in power stations.

Pellets are, by definition, a renewable. Since they are also cheap and easily transportable, they are very attractive to European generators, which must comply one way or another with the EU's renewables regulations. As luck would have it, they can be burned alongside coal in existing coal-fired plants for a relatively small conversion cost and with little or no loss of the coal's thermal efficiency. The more pellets generators can use in this 'cofiring' process, the more carbon credits they receive. In 2011, RWE Innogy, the renewables arm of the big German utility, commissioned the world's largest pellet factory in the US state of Georgia. Its aim was to supply fuel for cofiring at its existing coal-fired power stations in the Netherlands, and another one outside London, in the UK, that has been converted to burn biomass only.

Wood pellet production more than doubled between 2006 and 2010, according to the IEA.[38] The Biomass Energy Resource Centre of the US believes it will double again in the next five years, at least in North America. The product is not entirely clean, however. Although sulphur and nitrogen oxide emissions are low, wood pellets do release fine particulates that can be a problem in areas where many burners are clustered together.

I have been working with a company that is proposing a different twist on biomass energy. Its near-$1 billion project involves the world's fastest-growing tree, Trifolia, that the company developed itself over many years (without any genetic modification, as it happens). The highly cellulosic (therefore lots of energy) trees grow at a rate of eight metres a year, and then regrow on their own stumps. The wood will be fed into a pyrolysis or gasification plant, and the resulting gas used to generate electricity for sale to the local grid in the UK. According to initial projections, after the first three years the profit margin before interest and tax payments will be an extraordinary 80%.

Then there is waste-to-energy (WtE), which gives us the satisfaction of providing relatively clean energy while solving an unrelated problem. Instead of being thrown into landfill sites, which take up valuable land and may cause pollution problems of their own, domestic and industrial rubbish can be incinerated to produce heat and power. Interestingly, as we recycle more, the heat content of WtE's raw material, municipal solid waste, rises. That is because biogenic content such as newspapers and packaging produces less heat than nonbiogenic content like plastics. And as we recycle more paper and packaging products, the nonbiogenic share of our waste rises. The burning process produces some unpleasant emissions, such as particulates, acid gas, and traces of dioxin, but modern incinerators keep these to a minimum. Although burning waste reduces pressure on landfill sites, its critics say it reduces the urge to recycle or to keep waste to a minimum.

Waste incineration is the dominant WtE technology and a reasonably mature one. But there are other ways of extracting energy from waste, some new, some very well-established. Some thermal methods are still evolving, such as gasification and thermal depolymerisation. 'Plasma arc gasification' is a new WtE process currently meeting resistance from local environmentalists in California's Salinas Valley. End products can include gases such as syngas and hydrogen, or liquid fuels like synthetic crude oil. Certain non-thermal methods have been around for a long time but are now gaining much more traction than before. Fermentation can produce ethanol and hydrogen, and anaerobic digestion produces methane. These processes require biodegradable feedstocks, such as agricultural and food waste, or sewage. This is useful technology, but there is only so much waste in the world, which acts as a natural constraint on the amount of energy it is able to deliver.

In terms of overall lifetime cost of delivering power, biomass in general comes somewhere in the middle, as can be seen in the levelized cost of energy table. This takes into account all the costs of delivering energy over a fuel type's lifetime, including the costs of capital, fuel, operations, and maintenance, as well as the load factor and utilization rate – how often it is likely to be switched on and off.

Waste-to-energy does not get as many headlines as it deserves. Biofuels, which are also part of the biomass universe, get more attention than their size in the marketplace justifies – in 2010 they only accounted for 3% of transport fuel demand.[39] And yet the fact remains that only biomass is capable of providing alternative liquid fuels to the transportation sector. So, in the absence of great advances in fuel cell or storage technology, biofuels may become much more significant. Their story belongs in the next chapter, where we move from the world of power generation to the world of transport – a very different world indeed.

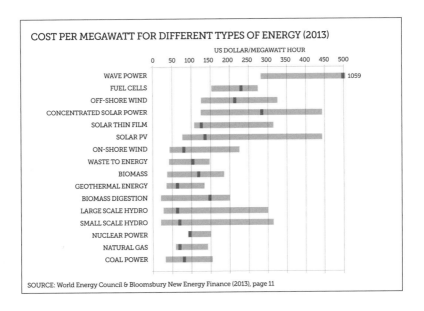

COST PER MEGAWATT FOR DIFFERENT TYPES OF ENERGY (2013)

SOURCE: World Energy Council & Bloomsbury New Energy Finance (2013), page 11

CHAPTER 3:

THE GREENING OF TRANSPORT

If we are addicted to the electricity that we devour at such a high carbon (and increasingly high financial) cost, we are no less hooked on our cars and our air travel. The transport sector consumes the lion's share of our liquid fuels – for which read oil – and is responsible for about one third of all man-made carbon emissions. Oil, as we have said before, manages to pack a lot of energy into a small space and, given the way that the car population and air travel are growing, it may prove to be a particularly stubborn addiction. There are some 800 million cars on the world's roads, and that number is forecast to double in little more than 20 years, by 2035.[40] As ever, much of the growth is coming from the east. China already manufactures more cars than the US and Japan put together and its own vehicle population could rise from today's 18 million to 30 million by the end of the decade. What is more, it is expected to expand its existing programme of subsidies to boost vehicle sales in rural areas. Those are all good reasons to look for ways to reduce transport emissions, and we are certainly not short of initiatives. But how successful will they be – and how soon?

With a vehicle population powered almost entirely by the internal combustion engine and fuelled by gasoline or diesel, the alternatives are fairly straightforward, at least on the surface. We could use our vehicles a lot less, as some environmentalists advocate, although you try telling that to a Chinese factory worker who has spent the past 10 years saving for his first car. Short of that, we either redesign the fuel, or revolutionize the engine. Let us start with the fuel. The first serious response to transport pollution has been to leave the internal combustion engine as it is, for the time being, and to develop combustible fuels, from non-fossil sources, that emit less carbon dioxide. These biofuels are derived from living things and are therefore a subset of biomass, which we looked at in the previous chapter.

Ironically enough, the first generation of biofuels has been as much a journey back in time as a leap into the future. The pioneers of motoring did not develop their internal combustion engines with petroleum products in mind. When Rudolf Diesel built the prototype of the engine that bears his name, he ran it on peanut oil. Henry Ford built his Model T believing that ethanol would be the fuel of the future. Then along came gasoline and petroleum diesel, more noxious but emphatically cheaper, and swept peanut oil and ethanol off the market for a century. But vegetable oils and alcohol are staging a comeback – or at least they were. The whole subject of biofuels has given us a sobering demonstration of the shifting political and social sands on which renewables are built.

Only a few years ago, biofuels were seen as an unalloyed 'good', a clean replacement for polluting petroleum products. So politicians introduced the regulation required to turn biofuels from a nice idea into a product with a guaranteed market. The certainty that regulations would force people to buy the product reduced the risk, and so encouraged investors and, up to a point, lenders to put money into the industry. Since 2005, the Renewable Fuel Standard (RFS) programme has required US fuel producers and importers to replace specific amounts of gasoline and diesel with biofuels. It stipulates absolute volumes rather than percentages, rising from nine billion gallons in 2008 to 15.2 billion in 2012 and 36 billion gallons by 2022. The EU has gone the percentage route, decreeing that biofuels must make up 10% of all transport fuel by 2020. As with renewables in general, some individual member states have their own targets beneath that umbrella, such as the UK, which has an intermediate goal of 4.75 % by 2014.

BIODIESEL

In Europe, the alternative fuel that has made most headway is biodiesel. It accounted for about 70% of the EU biofuels market in 2011, with bioethanol making up most of the rest. Prices for both have been falling, mainly due to competitive imports of biodiesel from Argentina and Indonesia (even though the EU remains the world's largest biodiesel producer), and of bioethanol from the US and Brazil. In the US, maize-based ethanol is the dominant biofuel, having enjoyed subsidies of up to $6 billion a year until Congress declined to renew them in 2012. Nonetheless, most US gasoline now contains about 10% ethanol, the maximum blend tolerated by standard engines before modification is required. US biodiesel usage is gradually growing, although off a lower base, with record production volumes of 1.1 billion gallons in 2011. China favours electric cars and its policy on biofuels is that they should not compete with food crops, so it has no biofuel subsidies or mandatory use targets. As a result, its production is about one tenth that of the EU's.

Biodiesel is made by 'transesterifying' vegetable oils or animal fats, and can be a constructive way of recycling waste such as used cooking oil. Feedstocks are many and various. Soybean oil is used widely in the US, whereas rapeseed oil is preferred in Europe. Other feedstocks include palm oil, tallow, and sunflower seed oil. The end product can be blended with petrodiesel or used on its own. Blends of up to 20% biodiesel (or B20, as they call it) can be used by standard diesel engines with little or no modification, although higher blends up to B100 may require more radical adjustments.

Biodiesel has also been used in trains. Since 2007, the Royal Train, used by the British royal family to travel around the UK, has run on 100% biodiesel made from used cooking oil. Prince Charles has well-tuned ecological sensibilities, so you may detect his influence in this greening of the family transport. Richard Branson's Virgin Trains experimented with a B20 mix in one of its UK express passenger trains at about the same time but then, sensitive to the gathering debate over the acceptability of first-generation biofuels, took it no further.

BIOETHANOL

Current gasoline substitutes are made in a different way, by fermenting and distilling sugars and starches, as you would do to make alcohol – which is what these products are. Ethanol is by far the most common product of this process. Another is butanol, which has a higher energy content than ethanol and can be used unblended in standard engines but, for the time being, is rather more expensive to make. The world's biggest manufacturers of ethanol are the US and Brazil, which together produced 70% of the world's total biofuels in 2011. Most of that was ethanol, made mainly from maize in the US and sugar cane in Brazil. The ethanol market is so established in Brazil that nearly all its cars run on E30 (30% ethanol) blends or higher.

Since all plants contain sugars, virtually any of them can be used to make ethanol – in theory. In practice, other commonly used feedstocks are sugar beet, wheat, sorghum, and cassava. But all of these commodities have another life as food, which has set off a backlash against the whole idea of biofuels. As more edible crops have been turned into fuel, there has been less available to eat, which has forced up food prices. The food price spikes of 2007 and 2008, and then 2010 and 2011, have been laid squarely at the door of these first-generation biofuels. What is more, as farmers have switched crops to serve the energy market – particularly noticeable in the US, where 40% of the maize crop now ends up in fuel tanks[41] – they have reduced the amount of land available to grow food. Large tracts of natural forest in South East Asia have been felled, releasing massive amounts of carbon dioxide and threatening the habitat of endangered species such as the Sumatran tiger, to make way for palm oil plantations. Other complaints are that making biofuels consumes a lot of energy, sometimes more than is present in the biofuel itself, as has been alleged for corn- and soybean-based fuels.[42] After calculating the energy involved in fertilizing, producing, harvesting, and processing, the energy input may exceed the energy output, leaving a negative energy balance. In the course of their life cycle, they may emit only marginally less, or even more greenhouse gases than fossil fuels. And, with the exception of Brazilian sugar cane ethanol, they are expensive to produce.

In 2012, the worst drought for 50 years savaged the US corn crop and intensified the critical spotlight on biofuels, now seen in many quarters to be taking the food out of people's mouths. As food prices have risen again and the public mood has changed, so has the political stance. Many Germans originally objected to biofuel quotas because they feared it might damage their cars. With their opposition now strengthened by food concerns, Germany's development minister felt able to call (although unsuccessfully, since he was serving in a coalition government) for sales of E10 ethanol blends at German petrol stations to be suspended.

The European Commission's response has been more far-reaching. Although its 10% green fuel target remains notionally in place, it has announced plans to limit *crop-based* biofuels to only 5% of transport fuel. At present, they already account for 4.5%, so there is not much room for increased production. Those who have invested on the basis of the original quota are dismayed, to put it mildly. "It's a big U-turn in EU policy," said Jean-Philippe Puig, CEO of French oilseed group Sofiproteol, when he heard the news. Sofiproteol owns the EU's largest biodiesel producer, so he has a keen interest in the matter. "We have made lots of investment in order to meet the 10% target in 2020, including more than €1 billion in biodiesel plants in France," Puig complained to Reuters. What politicians give, they can also take away.

The change in attitude is not confined to Europe. In August 2012, José Graziano da Silva, director general of the Food and Agricultural Organization of the UN, called on the US to suspend – temporarily but immediately – its biofuel quotas, so that more crops could be channelled into food and animal feed. In an election year, which that was, no presidential candidate with any sense would thumb his nose at Corn Belt votes by messing with ethanol. And having been re-elected, President Barak Obama appeared to be in no hurry to tinker with the quotas. But there are lessons here for advocates of wind and solar subsidy. One, as we have just seen, is that politicians are quite capable of changing their minds, so it is dangerous to build a future on the foundations of their promises alone. And another is that it is not unthinkable for green energy subsidies to prove to be misguided.

SECOND-GENERATION BIOFUELS

So the search is on for commercially viable 'second-generation' fuel sources, starting with plants that are inedible and can be grown on marginal or non-arable land too poor to produce food crops. There are quite a few contenders, such as jatropha, camelina, castor, and jojoba for biodiesel. Jatropha fuel, its champions say, burns with one fifth of the emissions of conventional fuel and gives back 20 times the energy required to make it. For ethanol, there are various 'cellulosic' grasses and woody materials such as switchgrass, miscanthus, wood chips, and agricultural leftovers such as straw. Because of their tough, lignocellulosic nature, these need to be broken down with acids or enzymes before processing, which adds costs and complications. Until now, second-generation fuels have been in the experimental stage, as would-be producers juggle with feedstocks, chemistry, and money. They have been trying various technology routes, some of which produce biochemicals and hydrogen as well as fuels. There is hydrolysis, which includes fermentation and the more experimental 'aqueous phase reforming', a sugar-to-gasoline process that is like oil refining but uses biomass as feedstock. By going the gasification route, biomass can produce syngas, which can then be processed into ethanol, diesel, or jet fuel. With pyrolysis, biomass is refined into liquid fuels after being heated to high temperatures in the absence of oxygen – so no actual burning.

Those who get it right, and can bring down the price, may grab a share of a market worth hundreds of billions of dollars, although we are only now beginning to see the early stirrings of commercial production. The world's first commercial-scale cellulosic (ie non-food) ethanol plant opened in October 2013 at Crescentino, on the banks of Italy's Po River. Beta Renewables – a €250 million joint venture between Gruppo Mossi & Ghisolfi, an Italian chemicals, plastics, and engineering group, and Texan private equity investor TPG Capital – produces ethanol there from a giant reed called *arundo donax*, as well as from wheat stalks and other agricultural residues.

Among all the feedstock options, there are particularly high hopes for humble algae, which, although at the small end of the scale, is still biomass and, as such, capable of producing fuel. It is also more versatile than most. Its lipids, or oily bits, can be extracted for biodiesel, and its sugars can be turned into ethanol or butanol. Methane can also be extracted from algae. An even greater attraction is that it can grow where few other crops would, not only on poor agricultural land but also in the sea or in waste waters. Best of all, as one of the fastest-growing plants on earth, it is considerably more productive than other feedstocks. Algenol, a Florida-based company that has developed a proprietary direct-to-ethanol process, says that it can produce 6,000 gallons per acre per year from algae, compared with corn ethanol's 370 gallons per acre. However, promising though it looks, algae may still be a decade or more away from being cost competitive.

Aircraft have been singled out as villains in the carbon emissions drama, their role intensified by a surge in cheap air travel over the past two decades. One early study puts the airline industry's share of global emissions at 3.5%, and they continue to grow faster than those in any other industry.[43] Airlines are pursuing increases in fuel efficiency as one way of reducing emissions, although they would be doing that anyway since, at today's prices, fuel represents their biggest cost. Greener fuel is another option and a number of ventures are developing fuels specifically for jet engines. With its nose for publicity, Virgin Atlantic was the first airline to go public with a biofuel flight. In February 2008, it flew a Boeing 747 from London Heathrow to Amsterdam, Schiphol using a 20% biofuel blend from coconut and babassu oil in one of its four engines. The flight was panned as a publicity stunt by anti-biofuel environmentalists and, as with the Virgin train trial, that seems to have been that for the time being. Richard Branson has been careful to say that future biojet fuel will probably be made from algae.

The International Air Transport Association (IATA) has told its members that biofuels must account for at least 10% of their consumption by 2017.[44] The EU has added to the pressure, insisting that, from 2012 onwards, airlines of any nationality using EU airports must 'reduce' emissions by participating in its emissions trading scheme. That obliges them to buy carbon credits to offset their emissions, adding to their costs and, as it happens, infuriating the US, whose taste for extra-territoriality fades rapidly when someone else is pursuing it.

Various players are jockeying for position in tomorrow's jet fuels market. One of the more visible is California-based Biojet Corporation, which calls itself a "supply chain integrator" for sustainable jet fuel. It has positioned itself with alliances and joint ventures all the way along the supply chain, controlling feedstock sources, developing refining capacity, and handling sales to end users. That is exactly what the oil majors do in their industry, but this is the first attempt to replicate the model in renewable fuels. Unlike most producers, it is feedstock-agnostic, with interests in jatropha, camelina, algae, and waste biomass. Biojet recently struck a deal with the 57 tribes represented by the Council of Energy Resource Tribes (CERT) to pursue feedstock, conversion, and refinery projects worth a possible $3 billion on Native

American lands – which offer considerable tax advantages. Biojet has recognized that, to make money out of early phase renewables, you have to track down every tax break and subsidy you can find.

And yet progress on biofuels, even the 'good' variety, is achingly slow. Many investors have lost money on jatropha projects, for example. D1 Oils, later renamed Neos Resources, was once the doyen of the sector. Refinery problems and – take note – the loss of a US biodiesel subsidy greatly diminished the company's prospects and, having raised more than £100 million from investors, it now appears to have ceased operations. Investors in many other quoted jatropha ventures have done just as badly. No-one is blaming jatropha science itself, and the belief is that it will, one day, do the business. But the heavy weather afflicting this and other biofuel feedstock initiatives demonstrates the unavoidable truth that, for alternative energies, the road to economic viability is very long and very hard.

Perhaps liquid fuels are not the answer. Perhaps even sustainable liquid fuels are only a slight improvement on a bad model. Perhaps the motor vehicle should switch to a completely different form of power, one that is, on the face of it, completely clean – electricity. Has the time finally come for the electric car?

THE ELECTRIC CAR

As it happens, this is another case of back to the future. Electric cars were commonplace in both Europe and the US during the late nineteenth and early 20th centuries, with good reason. You did not have to crank them to start them up, they had no gears to wrestle with, and, unlike the gasoline-powered competition, they were not noisy, smelly, or dirty. Electric cars, often steered with a tiller rather than a steering wheel, were popular with the rich. Electricity was also used to power trucks and buses over short, local routes just as, in the UK, it still drives a dwindling number of milk delivery carts today. In one sense, the electric car's undoing was electricity itself. The development of electric ignition in 1912 did away with the need for a crank-handle start. Now that Henry Ford's assembly-line, gasoline-powered cars were getting cheaper, that began to tilt the odds in gasoline's favour.

Just as low petrol prices once dimmed the attractions of the electric car, so high prices have brought them back into focus. But this time we also have the added spur of emission concerns. Electric vehicles had a curiously short-lived renaissance in the US during the 1990s, thanks to California, which was then suffering from the worst air quality in North America. The state's clean air agency, the California Air Resources Board (CARB), decreed that any manufacturer that wanted to sell cars in California would have to ensure 'zero emission vehicles' made up 10% of state auto sales by 2003. In response, General Motors launched the first mass-produced electric car, called the EV1. It had a lead acid battery and a range of 60 miles. Others followed, including Toyota, Honda, and Ford, even as those same car makers fought against zero emissions in the courts. CARB eventually backed off and GM recalled all its EV1s and destroyed them (tellingly, drivers had not been allowed to buy them but could only lease them). Writer/director Chris Paine tells the story in his very watchable 2006 documentary film, *Who Killed the Electric Car?*

Paine's line-up of suspects for the 'murder' included the following:

- the US consumer who, apart from a few zealous fans, was either unaware or uninterested
- the batteries, with their very limited range
- the oil companies, whose very business was threatened and who tried to obstruct the building of public battery-charging stations
- the car companies themselves, which did not want to disturb their existing business models
- the US government, with the powerful influence of the oil and automotive industries at work inside George W Bush's White House
- CARB.

Paine finds all of them 'guilty' to some degree, with the single exception of the batteries, even though they are linked to one of the film's more conspiratorial revelations. The batteries for EV1 were designed by a leading-edge firm called Ovonics. When Ovonics invented the nickel-metal hydride battery, which would have greatly extended the car's range, it was ordered by its majority shareholder not to tell anyone.

The majority shareholder was none other than General Motors. The car maker later sold its stake to another less-than-madly-supportive business – Chevron, the oil company that, later, was also accused of suppressing Ovonics's battery advances in its own interests. The long-suffering battery business was recently acquired by German chemicals giant BASF, which may prove a more sympathetic owner, although Stanford Ovshinsky, the extraordinary brain behind the company, died in 2012.

Whether you believe in the conspiracy theory or not, the film unerringly identifies the issues and the conflicting interests in the ongoing saga of the electric car. Big oil and, perhaps to a lesser extent nowadays, big auto have been wedded to the old technology. Although car makers are again venturing into electric cars, they would really like them to remain a minority interest – the fact that they require fewer new (and very profitable) parts and less servicing does not endear them to big manufacturers and their dealerships. The consumer still needs to be convinced. And the main reason why the consumer is not yet convinced lies at the heart of the problem. It is the battery – how to increase its capacity, and hence the vehicle's range, without making the battery so big and so heavy that it becomes impractical; how to make it quick and easy to recharge; and how to do all that at a cost that is not prohibitive. The storage problem – the solution to it is the alchemist's stone of electricity, whether it is for running a car or managing the grid. Everyone is searching for it, and whoever nails it will get very rich indeed.

A BETTER BATTERY?

One way is simply to build a better battery, except that it is far from simple. A battery has three important bits – the anode and the cathode (otherwise known as the electrodes) and the electrolyte. The electrodes are the conductors, the metal bits via which the electric charge enters or leaves the battery. As the current travels internally between anode and cathode it passes through the electrolyte, another conductor, which makes up the bulk of the battery. What goes on in a battery is a chemical reaction, which varies according to the nature of its constituent materials. Although electrolytes are a relatively simple affair, the substances used to make the electrodes are key. They determine the lifespan and the capacity of the battery and, most importantly, they also determine its cost. Having evolved through metals such as lead, zinc, and nickel to lithium and cobalt, they have been getting more expensive along the way. One important factor inhibiting sales of electric cars, although clearly not the only factor, is that they cost much more than their gasoline equivalents, and it is the cost of the battery that makes it that way.

The big car makers are finally embracing the electric car, some because they want to, others because they feel they have to. The Japanese, who love technology and futurism, have been early movers: Toyota with its Prius hybrid, and Nissan with its Leaf. In a convincing display of the power of the celebrity testimonial, as soon as American film star Brad Pitt bought a Prius, half of Hollywood put their names down for one. Carlos Goshn, who is CEO of Renault as well as CEO of Nissan, takes the electric car very seriously and Renault is now in the field with various models, including Fluence and the new Zoe. GM has acknowledged – reluctantly – that the future is finally arriving and is back out there, notably with its Volt hybrid and the all-electric Chevy Spark EV.

Early hybrids, such as Toyota's Prius, used nickel-metal hydride batteries. Most of today's electric cars use lighter, more compact, lithium ion batteries, which seem to work well in hybrids but less so in all-electric vehicles. Hybrids use an electric motor to supplement a gasoline engine, so use less fuel per mile and therefore emit less carbon. Their relatively small batteries are charged by a generator and by energy recovered from braking, and they cannot do much on battery power alone. Their high-power but low-energy batteries are good at accepting and delivering charge rapidly, but cannot propel the car at high speed or for long distances. That is not a problem for hybrids, whose gasoline engine takes care of the speed and distance. It is a bit more of a problem for plug-in hybrids, such as GM's Volt, which still have a gasoline engine but whose somewhat bigger electric battery can also be charged from the grid. And it is a big problem for all-electric vehicles, such as the Nissan Leaf, where the electric motor has to do all the work and the volume of energy stored is more important than the speed at which it is delivered. So these high-energy batteries need to hold much more charge, which they dispense slowly. The result is that they take many hours to recharge.

If it is the battery that makes electric cars so expensive, a number of governments are wheeling out the subsidy option to soften the blow. In Japan, China, the US, and many EU states, the cars trigger certain tax exemptions or cash rebates, or both. In Ontario, they get green number plates allowing them to travel in special carpool lanes. In the UK, in addition to an electric car grant of up to £5,000, the government will now pick up 75% of the cost of installing home charging facilities, to a limit of £1,000. If the idea is to subsidize until the electric car breaks even, a lot of our taxes are headed in that direction. It is up to us to decide if we think it worth the cost.

Opinion is divided about how much cheaper batteries can become. The car companies do not divulge the figures but today they cost perhaps $500-$600 per kilowatt hour, and can add $14,000 or more to the price of an all-electric car.[45] Renault's Carlos Ghosn believes that once electric cars are being manufactured in sufficiently large volumes, the unit price of batteries will come down enough to make them competitive with gasoline cars. That is assuming the electric car makers stay in business that long. Others point out that lithium ion batteries are already manufactured in large numbers for use in mobile phones and laptops, so much of the excess fat has already been squeezed out of the cost.

Either way, much research is being channelled into coming up with different, better batteries. Much of it concentrates on new electrodes – for example, electrodes based on the specially treated molecular-scale carbon known as carbon nanotubes. These batteries could be able to store twice as much energy as the present generation. Another possibility is a lithium air battery, where one electrode interacts with the air. It too would allow cars to drive much farther before having to recharge.

Carbon nanotubes are also being used in ultracapacitors, which have much longer life than batteries. That is because they store energy in an electric field rather than as a chemical reaction, so they do not degrade as fast. The ultracapacitor has been described as a small watering can with a wide spout – it is quick to fill up and empties with a powerful whoosh, but it does not hold much of the essential ingredient. High-energy batteries, by comparison, are like a big watering can with a very small spout. They can hold a decent volume of energy, which takes a long time to run out but also takes a long time to refill.

Some reckon that the battery's days in electric cars are numbered and that it will be supplanted by the ultracapacitor. One who feels this way is Elon Musk, co-founder of PayPal, now CEO of Tesla. California-based Tesla makes a range of all-electric cars, topped by its Roadster sports car (driven by the 'good' Ewing son in the latest TV series of *Dallas* – the 'bad' one drives a gas-guzzler). The cars use lithium ion batteries at present. If ultracapacitors are going to displace them, as Musk anticipates, their energy density will have to increase by a significant margin. It is possible that, instead of some winner-takes-all contest, batteries and ultracapacitors could be used in tandem, supplementing each other's strengths and weaknesses.

In the meantime, those who believe in batteries are trying to stay ahead, although the next big leap could be some time coming. Many of the more interesting battery chemistries are hard to scale up, depend on volatile materials, or are sensitive to changes in temperature. And the business conditions, as on any new technology frontier, are hostile.

All-electric cars do not actually solve the emissions problem. They only move it around. The electricity used to charge their batteries has to be generated somewhere, and it may well have been generated by a carbon-emitting, fossil fuel-powered plant. Fuel cells do get around that problem. A fuel cell is similar to a battery, with an anode, a cathode, and a liquid or solid electrolyte. Like a battery, it converts chemical energy into electricity by means of electro-chemical reactions. But whereas a battery contains a finite amount of 'fuel' that is eventually used up (and that is the end of the battery), a fuel cell is continuously supplied with fuel – usually hydrogen and oxygen – from an external source. This makes them very useful as standalone power sources in spacecraft, where they have been used for many years, or in hospitals, schools, and office buildings. Today, more attention is being given to the fuel cell's possible role in transportation. Car makers have been experimenting with fuel cells since the 1960s and, although no prototype passenger vehicle has ever been commercialized, forklift trucks powered by fuel cells have proved popular in warehouses where emissions are not wanted.

The US government has been offering incentives for hydrogen fuel research and development for the past decade, but the technology has a number of drawbacks. It is very costly, for starters. Size is also an issue, and fitting the equipment into a standard-sized car remains a challenge. Lack of a hydrogen-refuelling infrastructure is a major obstacle – no hydrogen pumps along the motorway. And safety remains a concern, since hydrogen is highly explosive.

Certain designs of fuel cell may lend themselves to large-scale power generation, although, once again, there are challenges of cost and reliability. And on the subject of power generation, storage could play a key role here too in the future. The management of the electricity grid, with its careful balancing of supply between peaks and troughs of demand, is bedevilled by the use-it-or-lose-it nature of electricity itself. The wind does not blow all the time, for example, and tends to blow strongest at night, when demand is low. If that excess could be stored for use when it is needed, it would improve the efficiency of the whole system and make it easier to integrate intermittent sources such as wind and solar into the grid.

A BIGGER BATTERY?

Batteries can be used, up to a point, to smooth supply and demand and provide short bursts of power, but they are not big enough to deal in huge, grid-scale quantities. Pumped storage schemes can do that, using cheap, surplus electricity when demand is low to pump water uphill into a reservoir. From there it can be released to flow downhill again, generating higher-priced electricity when demand is peaking. But that requires a hydro-electric scheme in the first place, and there are only so many places where you can build them. So recent research has been trying to replicate the pumped storage concept by other means.

In Denmark, for example, architects Gottlieb Paludan have come up with the idea of the Green Power Island. An artificial island is created in coastal waters, with wind and solar generators and a deep central reservoir. When demand is low, surplus energy is used to pump water out of the reservoir and into the sea. As demand rises, the seawater is let back in, driving turbines to generate new electricity.

Gravity is also at work in a scheme being devised by Advanced Rail Energy Storage (ARES) of Santa Monica, California, one of almost child-like simplicity. A series of parallel railway tracks lead up a hill, and special railway trucks with generators built into their bogeys are hauled up to the top, using off-peak energy. When demand rises, they are allowed to run down the hill, generating electricity. Its designers say that the scheme delivers more power than pumped storage for the same height differential. Gravity Power, another Californian enterprise, uses two connected vertical shafts, one wider than the other, and both filled with water. Water is pumped down through the smaller shaft to force up a piston in the bigger one. When the time is right, the piston is allowed to sink back down again, forcing water through a turbine. Unlike Ares, which needs a hill, this can be installed close to where it is needed, and extra modules can be added as required. Other designs store and redeliver compressed air, storing it in underground salt caverns, or heat.

Seen from an investment perspective, the one thing you can be sure of is that most of these technologies will fail commercially. What you can never be sure of, as you write out the cheque, is exactly which are the hallowed few with a viable future.

OR SOMETHING COMPLETELY DIFFERENT?

Perhaps the most ambitious project of all in this hyperactive field of storage has approached the car battery issue by proposing an entirely new infrastructure. In doing so, it also professes to offer a storage solution to grid operators. This is Better Place, the brainchild of Israeli-American software entrepreneur Shai Agassi. Having sold his business to German software multinational SAP, Agassi's hopes of becoming SAP's CEO were disappointed, so he turned his mind to the electric car. More specifically, he addressed the car's twin historic obstacles of cost and convenience. Electric cars must be 'refuelled', but how? One solution is a national network of charging points, at home, in the street, in parking garages, and in petrol stations. One company pursuing this route is the UK's Chargemaster, which is slowly rolling out a national network of its charging points. But as one newspaper story rather rudely pointed out, in some British towns there are now more charging points than electric cars.[46] It is a classic chicken and egg conundrum.

Better Place's Agassi had what he calls his "Aha! moment" when he hit upon the idea of separating the ownership of the battery from that of the car. In Agassi's model, the battery is not part of the car but fuel, and is paid for as a running cost, not part of the purchase cost. Obviously, the cost of the electric car minus the battery is substantially reduced. Instead, the proud new electric car owner signs up with Better Place, for membership that includes unlimited access to fully charged batteries at the company's 'switch stations'. Like petrol stations, these are (in theory) dotted around the country, providing the answer to the car's limited range. When the juice is getting low, you drop into a switch station and change batteries, which takes about five minutes. When you charge up your battery at home – which you will also be able to do – Better Place gets the bill. But Agassi did not stop there. He envisaged a nationwide network of switch stations and home-charging points all connected to and managed by a supercomputer, which is in turn connected to the grid. All those Better Place batteries in cars and switching stations, acting as one vast battery, could then serve as a buffer for the grid, storing surplus energy in periods of low demand and feeding it back during peaks. Phew!

It was a remarkable vision, and inspiring enough for Better Place to have won the support of the Israeli government, co-operation from Renault, $750 million in equity funds from investors such as General Electric, HSBC, and Denmark's Dong Energy, and a loan from the European Investment Bank. It rolled out switching networks in Israel and Denmark. And yet the vision was not inspiring enough to take the world along with it. Having burned through most of its cash, Better Place could only lay claim to 750 all-electric cars in Israel and Denmark combined. There was no money for further expansion until more people bought electric cars. But who would buy them until they knew a full switching network was available? Adding to the problem has been the hostility of some major car companies (although obviously not Renault), which believe that by giving away 'ownership' of the battery, they will be losing control of the whole automotive value chain.

The scale of the resulting tensions inside Better Place was revealed when Agassi stood down as the company's chief executive, although he remained a major shareholder and board member. Differences over future strategy were said to be the cause – Agassi wanted to consolidate Israel and Denmark first whereas others wanted rapid deployment into China and the US. But the company's low market penetration could not fund any form of network growth, and it finally filed for bankruptcy in 2013. Although the Better Place concept may be revived in one form or another, it remains a hard reminder that green dreams do not become the future overnight.

CHAPTER 4:

THE NEW ENERGY ENTREPRENEURS

Electric cars may or may not take off. One forecaster predicts worldwide annual sales of three million plug-in vehicles by 2020 (with nearly a third of them in Japan).[47] Although that would be only 3% of the light-duty vehicle market, the accumulating fleet would add considerably to the strains on electricity supply, which brings us back to the question of how we are going to put together our energy fix. We have seen which basic ingredients are available, so now we must create our drug cocktail. Let us think of ourselves sitting in a lab surrounded by powders and potions: a pinch of this, a spoonful of that, some drops of the other. In an ideal world we would simply use the cheapest, safest ingredients, but this is no ideal world. That powder is powerful, plentiful, and inexpensive, but it is also dangerous. This potion is relatively benign, but it is rare and costly. Even if a single ingredient was perfect for the job, it would be too risky to rely on it exclusively, in case we had problems getting new supplies in future. So we need a diverse mix of sources that are both obtainable and affordable. In the end, the recipe we draw up is dictated by our energy policy – in fact, the recipe *is* the energy policy. Now all we need are some chemists, or cooks.

To understand how the Green Bubble has been growing, it might help to take a closer look at the people and processes involved in the conception, funding, and construction of wind farms and solar farms, which are by far the majority of green energy projects. Bear in mind that none of this would be happening at all if it was not for deliberate energy policy – our recipe of favoured ingredients. Policy is where it starts. The government has promised its voters, or the EU or perhaps an international forum on climate change, that it will clean up its energy production. But it knows that no-one will invest in wind or solar energy as things stand, because there is no way that they could produce electricity for sale to the grid at anything close to the prevailing price. Grid parity is the Holy Grail for these emerging technologies, the point at which they no longer need to be subsidized to compete with fossil fuel-based electricity. Today, in general, although with partial exceptions in Spain and Greece, wind and solar are not able to compete unassisted. So the government does what governments do when they want to promote any activity that the private sector will not touch. They offer a carrot, for which we taxpayers and consumers ultimately pay. There is nothing wrong with that in principle, as long as our money is spent on a useful outcome.

The carrots are designed to motivate one target market above all others – the entrepreneur. I love entrepreneurs. They say money makes the world go round, but I think it is really entrepreneurs who make the world go round. These are people who will take more than ordinary risk if they believe there is a profit at the end of it, and without them we would still be living in the Stone Age. Some of them can be a bit too greedy for their own good, but this book is not an attack on entrepreneurs. Quite the contrary – they are what they are and they do what they do and, for the most part, we are all better off because of it.

The carrot that lures entrepreneurs out of their rabbit holes and into the green energy vegetable garden usually takes the form of feed-in tariffs. These guarantee a premium price for renewable electricity, a price that will turn an otherwise unviable project into a viable one. Germany was the first to come up with this particular incentive, introducing it in 1991 by means of the *Stromeinspeisungsgesetz*, the "power feed-in law". The Germans invented the 'feed-in' phrase as well as the idea and both have been adopted in more than 50 countries around the world. The German law did two important things: it obliged the grid to buy any electricity offered to it from renewable sources, and it decreed that the generators would be paid 90% of the retail electricity price. Since German retail rates are heavily inflated by consumption taxes, this was a good deal, worth considerably more than conventional generators were getting. However, it did have the disadvantage of fluctuating along with retail prices themselves. Since 2000 individual feed-in tariffs in Germany have been fixed, although they differ not only for every type of renewable energy but also for project size and numerous other variables. They are generally guaranteed for 20 years, although, as the years go by, tariffs for new projects come down according to a preset schedule. Even so, it has been a bonanza for green energy entrepreneurs and investors, who more than tripled German renewable output between 1999 and 2010. Power companies have to buy this expensive electricity, paying much more than they would normally be able to charge for it, and then German consumers pay the difference by having it added to their bills.

In some countries, alternative energy producers may also qualify for renewable energy certificates (see Chapter 2) that they can sell for extra revenue. Others use the tax system or grants to encourage green investment. Depending on the level of subsidy they offer, country by country, these carrots are of great interest to the private sector, as they are meant to be. There are different kinds of business people who are attracted to the green energy space, including some great entrepreneurs with strong track records, some from the power industry itself. But there are two types worthy of particular mention. One is familiar to us from the previous Bubble – the former real-estate developer who cannot resist the lure of risk-free money.

Many solar and wind projects, particularly at the smaller end of the scale, start with a local developer. In one very real sense, this Green Bubble is the child of the last great Bubble, the Property Bubble, whose collapse unleashed the financial crisis. During the boom years, when credit was freely available, property was everyone's favourite asset. Acquiring it was so easy – banks would lend to just about anyone who wanted to get into the real-estate business. In Spain, for example, they would lend you up to 120% of the property value, so you not only got the property but also a cheque in your hand. As long as banks kept on lending and property prices kept rising, even a fool could make money and, not surprisingly, the game attracted a lot of self-starting, business-minded entrepreneurs. However, then came the credit crunch and the game blew up. Since no-one was lending money to anybody, the property business shrank overnight. But property entrepreneurs needed something to do and, much to their relief, along came green energy.

Here, ripe for development, was a different type of project but with one important similarity. Wind farms and solar farms both start with a piece of land. The property entrepreneurs might have lost their old business model but they retained one very pertinent skill. They understood the property value curve and how to advance it from an empty piece of land, worth X, to a piece of land with a structure on it, or at least permission to build a structure, worth X plus Y.

Our entrepreneur – let us call him Josef – has local roots, which are key to what comes next. With his business history, he is well-acquainted with the properties in the area. He knows where some likely land for a solar or wind project can be found, owned by one or more of the local farmers. He even knows the farmer, or his brother-in-law does, or he once did business with the farmer's cousin, so when he calls to suggest a coffee and a chat he is not a complete stranger. Josef does some sums on his spreadsheet, just like in the old days, computing costs and possible returns, so he has worked out just how much he can offer the farmer for his land and still make a profit.

For the farmer, it is a question of comparing how much Josef is offering with what he can earn from planting the most profitable crops or pocketing set-aside payments. The entrepreneur might offer to buy the land, but he is more likely to suggest a long-term lease for, say, 50 years. At this stage, he will probably pay only a small sum for an option on the land. He does not want to run up his costs until he knows he has pulled together a complete package.

Once Josef knows he has the land lined up, he needs various permits and permissions if he is to create a saleable package. Here too his local connections come into play. He is well-acquainted with the municipal planning labyrinth. He has had dealings with the appropriate officials and politicians in the past and possesses the patience and thick skin required to see the planning application through to a positive conclusion. It is a detailed and time-consuming exercise and, in some countries, having the right people on his side may be vital. Josef will have to obtain a variety of consents and jump through all sorts of different hoops. One very important piece of paper he needs is a grid connection agreement. In remoter areas, where transmission lines do not yet exist, physical connection to the grid could be another considerable challenge. He may also have to carry out a number of technical and environmental studies. If it is a wind farm, for example, he might need clearance from the aviation authorities, which worry about large structures sticking up into the air. This all costs money and there will be fees to accompany the applications. By the time Josef's project is ready to go, with all the necessary permits, he may have laid out anything from €100,000 to €500,000 of his own money, perhaps even some millions of euros, depending on the project's size. For our purposes, we will say that the land Josef has identified can support a 70-megawatt wind farm. That is big for an independent developer, but it will make our sums easier. Because of its size, Josef has had to lay out about €2 million, which he has raised from his own resources and from family and business associates. No bank would dream of lending him the money at this stage. There is always the risk that permissions are refused, but that is a risk Josef is prepared to take in the hope of ample reward. Whether it is a solar or a wind project, the application process will follow a similar trajectory. If and when all the permits are in place and the land deal is finally closed, the project enjoys what bankers would call its first 'value uplift'. Suddenly it is worth a whole lot more, so now it is time for stage two.

Stage two is going to cost real money. It is just possible that Josef will undertake stage two himself, but it is unlikely. That would require investment on a scale that is almost certainly out of his reach and, anyway, building solar or wind farms is not what he does. He would rather realize the value he has added, and flip the project over to someone who does build solar farms. He prefers to take his profit and get on with assembling another package while market conditions are still favourable. So, exit the entrepreneur and enter the developer.

It is at this point that bankers like me will start to get sight of the project – and I have seen many of them. Now, let us say that our Josef, who does not really exist, sells the project – essentially the land rights and permits – to Greenzone. Greenzone does exist, and I have worked with it on a number of green energy projects. Although it employs a number of people, its proprietor and its brains is its general director, Frank de Wit. Frank is a Dutchman, based in Oslo, who cut his teeth in the telecoms industry, where he specialized in international project development and finance. The experience he picked up there – particularly in finance – is integral to his present career as a green energy entrepreneur. Greenzone agrees to pay Josef €7 million, which it will hand over in instalments, just to make sure the developer is still on hand if anything goes wrong.

Solar and wind projects start to diverge at this juncture and to follow slightly different routes. Wind projects demand more in the way of time, engineering skills, and money. The equipment is more expensive and, with order bottlenecks in the system, can take much longer to be delivered. All in all, the whole business is likely to take rather longer to come to fruition – possibly a couple of years. CSP solar is, likewise, a major engineering project that cannot be done in a hurry. Solar PV farms, on the other hand, are relatively quick and simple to build and can be completed in months rather than years. Frank specializes in taking solar panel projects from green light through to physical completion. That means sourcing the equipment, organizing the construction of the plant, and, most importantly, lining up the finance. He has to do that in such a way that he makes a decent profit for himself while leaving enough meat on the bone for the next purchaser when he sells it on – which he fully intends to do.

Let us imagine that getting the solar farm to the point where it is plugged in and generating is going to cost €100 million, including the fee that Greenzone will pay to itself. That might break down roughly as follows:

	€ million
Land	7
Professional services (lawyers, engineers, and so on)	4
Equipment (panels, transformers, and so on)	45
Construction	35
Greenzone's margin	9
TOTAL	100

Frank would borrow it all if he could, but no bank will lend to a capital project that is financed by 100% debt. They want to see the borrower stump up 20% or 30% of his own cash or 'equity', as they call it, and, as long as they think it is not too risky, they will then lend the other 70% or 80%. In this case, Greenzone and various of Frank's associates and partners are able to come up with €25 million in equity. Now they must borrow the rest.

Debt is vital to the project's success. With 100% equity, it could never produce a high enough return to justify the investment. This is the principle of leverage at work. Leverage states that, if all goes well, the smaller the equity portion, the higher the returns. Leverage is a fancy word for debt, and is sometimes called 'gearing' because, like a gear wheel, it magnifies returns. If things go badly, however, gearing also magnifies the losses.

If you invest €100 in a venture that makes a €20 profit for you, that is a 20% return on your money. Not bad. But what if you only invested €25 of your own money and borrowed the other €75 at, say, 10% interest? You still have a gross profit of €20. After paying €7.50 in interest, you are left with €12.50 in net profit, which is a more impressive 50% return on your original €25. And you still have €75 that you could invest in another three projects on similar terms. That is the dangerous magic of leverage. If the venture lost €20, however, it would work the other way. If you put up the whole €100, you would have lost 20% of your stake. If you only invested €25 in cash and borrowed the rest, after paying €7.50 in interest you would be down by €27.50. Your losses are now a scary 110%. Leverage is Bubbly. It helps to inflate Bubbles, but it also ensures that, when they burst, the damage is gruesome.

Frank needs debt financing partly because he does not have €100 million but also because, with the risks he is taking, he wants to leverage the returns to make it all worth his while. Banks are inclined to lend to Frank, or rather to Greenzone, because he has done this before and they know he can 'execute'. It is a green project, which is a plus. It flatters the bank's profile up in the corridors of power as well as down in the high street to be associated with 'green' these days. But what will really swing it for the bank is if Frank can produce a power purchase agreement to prove he has a solid buyer for the electricity at the elevated feed-in tariff for the next 20 years. If the tariff is guaranteed by the government, it means that the bank's counterparty is not really Greenzone any more but the government. And that makes it feel much safer about lending the money. The utility that buys the electricity pays the premium and it will pass on the cost to its customers – via a green levy, perhaps, or by the less transparent means of spreading the extra across its entire pricing structure. As an alternative to this type of off-take agreement, the promise of renewable energy certificates is of some comfort to a lender but, since their value is likely to fluctuate, this is not nearly as bankable as 20 years of guaranteed feed-in tariff.

Frank has another card up his sleeve, one that will help to make his sums look even more attractive. He is importing his solar panels from a manufacturer in Norway. They are better quality than the Chinese version, which is good, but they also cost more, which is not so good. However, as the customer of a Norwegian exporter, Greenzone can benefit from an export credit guarantee. This means that Norway's state-owned export credit agency will guarantee up to 85% of the debt portion of its financing. In other words, it will stand surety for a sizeable loan from Greenzone's chosen bank. Since the lending bank's counterparty is no longer Greenzone but effectively the Norwegian government, the interest rate on the loan is much lower than it would otherwise be. Frank has to borrow the balance elsewhere at market rates, but the effect on his all-in rate of interest is significant, lowering it from about 9% to perhaps 5%. (Interest on the Norwegian-guaranteed portion may be as little as 3%.)

Once he has organized his financing, Frank hires a so-called 'EPC' (engineering, procurement, and construction) contractor to build the plant. Time is of the essence, because the solar farm does not lock in its preferential tariff until it is plugged into the grid. Recently, governments have been cutting tariffs without notice and with increasing frequency, and if that happens before Frank has his revenues cast in stone, his sums may no longer work. If he succeeds, Frank will lock in a price of €0.25/kWh. That is more than four times as much as the standard rate of €0.06/kWh, which illustrates just how much extra cost electricity consumers will have to absorb in the name of government policy.

It is a cat and mouse game, especially in solar. The government's aim is to encourage green energy generation, not to make entrepreneurs needlessly rich. As it is perfectly aware, the total costs of building a wind or solar farm have been falling as the months and years go by. There are two distinct reasons for this. One is because technological advances are increasing the efficiency of solar panels, which are now delivering more energy for the same price. So project developers can use fewer panels to deliver the same amount of energy (which means lower costs) or use the same number of panels to deliver more energy (and earn higher revenues). The other, more dramatic reason is the appearance on the scene of the Chinese. China has pulled out all the stops to become a major player in renewable energy. It introduced a range of subsidies, tax breaks, and cheap loans to encourage its domestic industry, and grew to dominate international renewables manufacturing in only a few short years. Although China has brought little in the way of technology to the party, its cost advantages and manufacturing volumes have brought equipment prices crashing down, most notably in solar panels but also in wind turbines. Chinese oversupply has added further to this downwards pressure on prices, creating troubles for Western manufacturers and triggering trade disputes with both the US and the EU.

As project costs have fallen and efficiencies have risen, so governments have been lowering feed-in tariffs to try to keep the margins on offer within respectable limits, and the trend for feed-in tariffs is inescapably down. When a tariff is cut, it may make projects uneconomic in country A, so the green energy entrepreneurs will move on to countries B and C where the sums still work. In their Bubbly way, they do not care which country they are operating in – they just need the sums to work. Then panel costs come down still further, or technology improvements add to efficiencies, which makes doing business in country A feasible once again. At some point, the government will decide that it has all the green energy it needs, or that the technology is viable on its own, and will no longer offer any sweeteners at all. That is the theory, anyway.

The point is that Frank needs to switch on his new solar farm as soon as he can, before the government moves the goalposts. By the time he does, he has created a lot of extra value. The plant is now a cash machine, producing revenues every day, revenues that can be reliably predicted for the next 20 years. That makes the business easy to value. Like Josef, Frank could conceivably hang on to the project. It is now a very attractive asset, very like a bond, with a predictable and high yield. Its profit and loss account might look, very approximately, something like this:

Revenues	(€ million)
84m kWh @ €0.25/kWh	21.0
Operating expenses (including operation and maintenance, administration, insurance etc)	1.6
Depreciation	6.0
Interest	4.6
Guarantees, agency fees etc	4.3
Taxes	0.5

Total deductions	17.0
Net income	4.0

That is a 19% return on revenues after taxes, which would make the business attractive to potential purchasers, depending, of course, on the price they paid for it. Note how little the plant actually costs to run – the fuel is free and the staff requirements are minimal (although operation and maintenance costs would be somewhat higher for wind). But note also that if the electricity were sold at the standard tariff of six cents per kilowatt-hour instead of 25, it would not even begin to work, since revenues would be little more than €5 million (84m x €0.06 = €5.04m) against total costs of €16.5 million before tax.

Frank cares about return on revenues, but he cares even more about his own return on investment. Against the €25 million equity that he and his partners put in, the income represents a 16% return. But the real return on his investment will be dictated by how much the plant's new owner will pay for it. Who will it be? Frank's operational solar farm may attract a number of different buyers. One could be a large electricity utility. Big utilities have a problem, created for them by the government. They produce most of our electricity, largely from fossil fuels. Now they have been told that for every megawatt they produce, they must have a set number of green certificates – in other words, they have to buy the right to pollute. So what do they do? The most obvious solution would be to build their own solar and wind farms, generate renewable energy, and earn their own certificates directly. Some of them do. Dong, the Danish utility, has built a lot of offshore wind capacity, for example, but then offshore wind can be done on a meaningful scale.

The problem is that these are very large companies with very large balance sheets. When they build new generating capacity they are used to thinking in gigawatts, and big, for them, is beautiful. Green is not so beautiful. From their point of view, not only are the individual project capacities often tiny, but putting together a wind farm is also very labour intensive – having tea with the quirky farmer and his neighbour, assuring them that the birds are going to be okay, dealing with the municipality. And all of this for what, to them, is a mere rounding error in terms of finance and incremental capacity. However, if someone else is prepared to do all that fiddly work, they might be interested in buying the result, especially if they can bundle together a number of small wind or solar farms in the same region. The green energy regime may be a nuisance to utilities, punishing them for their legacy fossil fuel-based generation, but they benefit overall, since it is a good excuse for them to raise prices in a less transparent way than their regulators will usually allow.

Pension funds are also potential buyers. Pension funds take in regular contributions from their members to be invested in assets, grown (they hope) in value, and eventually returned to their members in the form of pensions. Since they really do not want to lose their members' money, they are very conservative investors, constantly on the lookout for safe bets. But safe bets do not pay very well these days. The flight to quality in the wake of the financial crisis has pushed up the prices of the least risky government and blue chip corporate bonds and, as a result, forced down their yields. The fund may be getting 2% or 3% from its bond portfolio, when what it really needs to service its liabilities – to pay out to its pensioners – is more like 4% to 5%. One way of boosting their returns, pension funds have discovered, is to invest in 'infrastructure'.

Infrastructure assets might be airports or toll roads, water companies, or, indeed, power generators. What they have in common are stable, predictable, and inflation-linked long-term cashflows that are measurable for 10, 20, or more years into the future. The longevity of the asset is important to pension funds and life insurance companies as they become increasingly keen on 'asset liability matching' – matching the long-term value of their assets to the long-term value of their liabilities. Infrastructure assets also enjoy a predictability and sustainability of income that meshes with the funds' low appetite for risk. This could be because they are monopolies or because their business is protected by high barriers to entry. Perhaps they hold long-term concessions, as in a toll road, or off-take contracts that guarantee income for an extended period. So a solar farm with a 20-year power purchase agreement certainly fits that bill. If Frank sold the project to a pension fund for €50 million, he and his associates would have doubled their money, and the fund would enjoy a superior (for them) return of 8% from an asset that is very like a bond. It is riskier – there could be operational troubles or problems with the panels, resulting in less electricity and less income. For that reason, the pension fund is not going to sink too much of its money into alternative energy. But the risk is tolerable and, if nothing goes wrong, the yield will spice up the fund's returns overall.

A few years ago, a pension fund that was interested in this kind of asset would have taken the safer route of investing via an infrastructure fund. These investment funds are specially designed to invest only in infrastructure-related companies and schemes, but they diversify their risk across many different projects. Now some of the bigger pension funds and insurers feel confident enough to invest directly in infrastructure projects themselves. On that basis, the Ontario Teachers' Pension Plan in Canada now directly owns two marine container terminals, one in New York City and another in New Jersey, as well as 39% of Brussels Airport and 50% of an electricity transmission and distribution company in Chile.

There is a third category of potential purchaser for Frank's solar farm – what you might call financial investors. They include the aforementioned specialized investment funds, which might have an infrastructure theme or a green theme, or an energy theme. They too might be happy with an 8% yield in these straitened times. If and when it finds a new owner, the solar farm can be metaphorically locked away in a safe, where it will generate a nice, predictable income for the next couple of decades, based on generous electricity prices that consumers will be paying for. (As long as the government does not renege on the deal – but we will hear more on that subject later.)

Today, the average pension fund holds less than 1% of its total assets in infrastructure investments, and the merest fraction of 1% in renewables, according to the OECD. But more are becoming alert to their attractions. PensionDanmark, one of the larger Danish pension funds, has already invested more than 6% of its portfolio in renewables and wants to raise that to 10% within a few years. Its holdings include a stake in Denmark's Nysted offshore wind project, one of the world's largest, and it is a significant bondholder in Sweden's Jadraas project, the largest onshore wind farm in northern Europe. Munich Re, the German reinsurance company, plans to invest €2.5 billion in renewable energy assets by 2016. German insurer Allianz has promised to add to the €1 billion in renewable assets that it already owns.

So that is the story of Josef and Frank. A variation on the Josef-entrepreneur model is provided by Oamec Energy chairman Douglas Gardner, a real-life solar park developer who has turbo-charged his schemes not with feed-in tariffs but with tax breaks. Gardner is a qualified actuary who spent much of his career in asset management. Green energy may be relatively new to him but investment returns and taxation are matters he has always understood. Attempting to assemble a solar package in the UK with the use of ROCs (Renewables Obligation Certificates, a UK form of green credit), he could never push the returns higher than 7%. The potential investors he spoke to felt this was not a high enough yield for the risk involved. Then he tried wrapping the investment in the UK's Enterprise Investment Scheme (EIS). This gives 30% tax relief on investments of up to £1 million in qualifying small companies, as long as the shares are held for at least three years. Gardner's plan is to sell the solar farm after those three years, at which point he will return £1.25 for every £1 invested. Nominally, that represents a compound annual return of just over 7%. But because of their EIS tax break, the investors will actually have invested only £0.70 – which means they will enjoy a vastly improved return of 21.5%. And their gains are tax free, making them worth even more in the real world. Selling the project will not be difficult, he says, and there is effectively a list price per megawatt, which is why he can predict the returns. The hard part is finding suitable sites to develop. Because the EIS rules limit the size of each company, Gardner wants to do 25 separate solar schemes.

Gardner's structure is pretty straightforward stuff. However, the UK authorities' religious zeal for green energy has led to packages where the combination of different reliefs and subsidies is generous to the point of madness. I have been looking at financing the purchase of a large US biofuel plant by a UK renewable fuels company. With lots of risk-free government money, it is worth the company's while to dismantle the plant and ship it piece by piece back to Britain. Meanwhile, a tax reduction specialist hedge fund has designed a structure whereby high-net-worth individuals can invest up to £1 million in return for shares in this company. Shortly thereafter, the investor gets back 85% of his investment in tax relief. So he now owns shares for which he has only paid 15% of the value. Nice. But it gets better. The company's revenues are, as ever, supported by a government subsidy of one kind or another, which means that you and I as taxpayers are paying close to 100% of the investment money. And if, in spite of all this free money and free risk, the whole enterprise should fail, there is further tax relief available, which can actually take total relief above 100% of the original investment. These opportunities are, naturally, only available to the very rich. How sustainable is that?

There is another noteworthy entrepreneurial type trying to make things happen in the new energy market, but one with a very different motivation. This person is not driven principally by financial modelling or economics, but by technology and the dream of "disruptive innovation". The phrase was coined by Harvard business professor Clayton Christensen in the late 1990s to describe innovation that creates a new market, and sends all existing manufacturers or providers back to the drawing board. The personal computer, budget airlines, Amazon, and Skype were all market disruptors. Would-be disruptive entrepreneurs usually come out of the world of technology, so they understand it. They may also have been very successful at it, with ample funds that they can sink into development of the new idea. Although they may not be primarily concerned with the financial sustainability of the innovation, they are well aware that, if it succeeds, the revenue potential is enormous. They have a sense of being on a new frontier, on the verge of something that will both benefit people and generate loads of money.

The idea of saving the planet and getting dazzlingly rich at the same time has its own powerful magnetism so, although alternative energy is not the only sector that attracts aspiring disruptors, it gets more than most. One of the most visionary examples of the breed is Shai Agassi, founder of Better Place (see Chapter 3). Another example is Elon Musk, the South African-born cofounder of what became PayPal, the online payments system. The fact that Musk has University of Pennsylvania degrees in both physics and economics is a clue to what turns him on. He says that when he graduated there were three areas that interested him: the Internet, clean energy, and space. The sale of PayPal, an online payments system that is very much a creature of the Internet, made him a multi-millionaire, giving him the wherewithal to pursue the other two.

Musk founded SpaceX, a space transport company that has already sent a cargo-laden capsule to the International Space Station and hopes one day to send humans to Mars. That is of great personal interest to me, because I have signed up for three space flights. But today he is also chairman, chief executive, and, as he insists on being called, 'product architect' of Tesla Motors, the New York Stock Exchange-listed maker of all-electric sports cars. He did not found the company but put up most of its early finance and gradually came to dominate it. The company has spent more and taken longer than anticipated – one virtue of rich visionaries is that they keep writing cheques long after everyone else would have folded their tent – but it has sold more than 2,350 of its limited edition Roadster sports cars (body by Lotus and 0-60mph in under four seconds). Its Model S luxury sedan has reportedly sold upwards of 25,000 and its Model X, SUV-meets-minivan, is scheduled for launch in 2015. It sells electric powertrains to Daimler and Toyota, both of whom are minority shareholders in Tesla. The Roadster is definitely a niche product, but the other models have set their sights on a mass market. So far, so good but, like all disruptive technology, only time will tell. In the meantime, Musk is still allowing his imagination to explore the future – he thinks that one day we may have flying cars.

CHAPTER 5:

IT'S NOT EASY BEING GREEN

Sunning itself at the heart of Bubble psychology is an uncritical, almost irrational belief in the absolute and irreversible benefits of a trend or an idea. In the mid- to late-1990s, investors were set on fire by the idea of the internet. The Internet Bubble began in earnest in 1995 with the stock market flotation (or IPO – initial public offer – in tradespeak) of Netscape, the US internet browser business. There were two striking things about this listing. One was that so many investors wanted a piece of the action that, at the very last minute, the issue price was doubled from $14 per share to $28 (it hit $75 on the first day of trading before gradually returning to earth). The other was that all of this hysteria was generated by a company that had never, not ever, made a profit. The pattern would be repeated many times as other unprofitable dot.com businesses – some with no revenues, never mind profits – were euphorically welcomed to market by investors. Many of them simply faded away, including Netscape, although its name survives as a run-of-the-mill internet service provider in the AOL stable. Having paid the equivalent of $10 billion in its own shares to acquire Netscape, AOL became the Internet Bubble's most spectacularly bedazzled victim.

Although I describe what is happening in the green energy market as a Bubble, purists could argue that I am misusing the word. Technically speaking, a Bubble occurs when asset prices become over-inflated, and the assets in question are most often shares or property. Then it bursts and investors are hurt, particularly those who arrived at the party last and paid the most. The first one we remember was not called a Bubble but a 'mania' – Tulip Mania – which is perhaps a more accurate description. In well-to-do 16th century Holland, tulips were newly arrived from the east and quite different from any flower seen before. They became so sought-after that prices reached considerable heights. But it was only after speculators entered the market that the frenzy really took hold. The form of paper trading that evolved became known as *windhandel* or 'wind trade' because no bulbs were actually changing hands any more. At the mania's zenith, in 1637, a single bulb is said to have changed hands for five hectares of land, shortly before the market crashed back to earth.

'Bubble' entered the lexicon in 1720, along with Great Britain's South Sea Bubble. The South Sea Company was fraudulent from the start, in the sense that its real purpose was to operate as a bank. But the Bank of England, founded some years earlier, was the only joint-stock company allowed to call itself a bank, so the new company pretended its mission was to compete with Spain in South American trade. Investors did not seem to care much what the company did, but they had to have the stock, which traded up from £128 for 100 shares in January 1720 to a dizzy peak of £1,110 in August. That valued the company at 10 times the national debt. By December, after everyone finally realized that this was unsustainable and started racing for the exit, the price had plunged back to £124, leaving much bankruptcy, suicide, and pain in its wake. Thereafter, Bubbles grew and burst with increasing regularity. When Wall Street crashed in 1929, after a particularly breezy Bubble, it was only the latest of a dozen such broader market 'panics' in the preceding 100 years.

The Internet Bubble, which finally burst in 2000, was essentially another stock market bubble, based on the idea that the world wide web was the future and that getting in on the ground floor was a sure way to grow rich. In its way, it *was* the future, but investors misjudged how long it would take to bed down in the marketplace and were too eager to back unsustainable business models. Those infected with irrational exuberance were mainly the dot.com entrepreneurs themselves, investors and investment bankers – some more cynically than others, as later became apparent. They were all egged on by the media, as opportunistically selective as ever in the application of its critical faculties.

The Green Bubble is less hysterical but much more broadly based, which actually makes it even more dangerous. It is not just, or even mainly, about getting rich, although smart green investors have been lining their pockets at our expense. This Bubble is being inflated by a passion even more powerful than greed – religion. Not 'church' religion, of course, but a belief system, nonetheless, in which we are convinced we have to do certain things if we are to save – to 'redeem' – our planet. Indeed, at the extremes, there is a whiff of evangelistic zealotry about the green movement that is reminiscent of the more revivalist religious sects. As well as casting its glow over the usual suspects – entrepreneurs, investors, investment bankers, media — the halo effect that surrounds anything 'green' has also beguiled the public at large, commercial banks, politicians at all levels of government, the European Commission, and various international and supranational forums.

This may not be a strictly classic Bubble in the South Sea tradition, but it is about the inflated, ultimately unsustainable values of assets – businesses in this case, built on considerable amounts of debt. The late Massachusetts Institute of Technology economics professor Charles Kindleberger, author of *Manias, Panics, and Crashes*, used to say that Bubbles typically occur after markets get unexpected good news. Greed kicks in and is then overdone. It is a fundamental truth that markets are fuelled by fear and greed, and that their direction is determined by the extent to which one outweighs the other. But they are equally powerful, and the green energy phenomenon is different only in that it had its roots in fear rather than greed. Now we have morphed our fear of climate catastrophe into green energy action and are marching greedily forward with 'good news' in our hearts.

The Green Bubble has built on the foundations laid by the Club of Rome in its 1972 publication, *The Limits to Growth*. This told us that, in a world where resources were finite, it was not rational to expect continual growth. That is a powerful argument, but politicians do not so readily delete 'growth' from their rhetorical arsenals. So to advocate the greening of our economies has been a god-sent way for them to square the circle, by standing up for ecological responsibility *and* growth. Growth and, above all, green jobs. The EU has promised us that cutting emissions and investing in green energy will create half a million new jobs. EU parliamentarian and, later, Swedish environment minister Lena Ek went some steps further and promised six million new jobs if only the EU would raise its 2020 carbon reduction targets from 20% to 30%. In Britain alone, the Carbon Trust has predicted 250,000 new wind industry jobs by 2050, and US President Barack Obama promised five million, if only you would vote for him and his energy policies.[48] As we will see later, green energy's inflationary effect on electricity prices has actually destroyed jobs in some countries, by hurting energy-intensive industries, but to mention it is like farting in church. We must not say uncomplimentary things about green energy.

During the Internet Bubble, if a venture was dot.com, part of the much-trumpeted 'new economy', it was okay. As long as a business could show it had a certain number of clicks per user, the usual valuation methods did not apply. Today, if it is green, it is okay. We will promote it, subsidize it, legislate for it, lend to it, and invest in it. Since it is for the ultimate benefit of the planet, it a) is good for us (the bank, the government, and so on) to be seen doing it and b) somehow demands less critical scrutiny than other, non-green projects. Even if you have your doubts, it is prudent not to express them, lest you seem an enemy of progress and goodness, a self-serving champion of the wicked, filthy old ways of doing things, for whom the life of the planet weighs less in the balance than your own shallow comfort and convenience.

If the believers back the idea of using our taxes to encourage green energy, they are energetically supported by opportunist entrepreneurs who can see the potential for profit. Together they form a powerful lobby. There is a phrase in academic circles for this kind of political coalition between missionaries and carpetbaggers – they call them 'baptists and bootleggers'. This evocative phrase describes a political model, first articulated by economics professor Bruce Yandle, in which opposite moral positions produce the same vote. In 19th century America, laws banning the sale of alcohol on Sunday were supported by both baptists and bootleggers, the former for moral reasons and the latter because they suppressed the competition. Politicians who backed keeping the laws on the statute book could claim to act in the name of public morality, even as they accepted campaign contributions from bootleggers.

In 2001, Yandle co-wrote a paper entitled *"Bootleggers, Baptists, and the Global Warming Battle"*[49], claiming that comparable influences were at work in the Kyoto Protocol. As the article's authors explained: "We argue that a similar phenomenon took place in the battle over the Kyoto Protocol, where the 'Baptist' environmental groups provided moral support while 'Bootlegger' corporations and nations worked in the background to seek economic advantages over their rivals."

Today's public sponsorship of green energy gets similarly bi-partisan backing from both baptists and bootleggers. Warren Buffett, the Oracle of Omaha, is now the second-largest solar operator in America via his utility subsidiary, MidAmerican Energy Holdings. The company also has interests in wind, hydro, and geothermal. But Buffett is a bootlegger, not a baptist. Note that he is a complete technophobe, and almost never goes near technology stocks. He is much more comfortable with established, easy-to-understand brands such as Coca-Cola and Gillette. For him to invest in an area where the technology is still evolving means it is just too good to miss. Because he is the Sage, the attractions of an investment guaranteed to generate large cash flows for the next 20 years are not lost on him. Jeff Siegel is an unabashed bootlegger. As managing editor of *Power Portfolio*, a US newsletter that ostentatiously promotes investment in renewable energy, this is what Siegel tells would-be investors: "The next time you look at a solar panel, wind farm, or state-of-the-art, super-efficient cogeneration plant, remember one thing: These were not built for treehuggers; these were built for very wealthy investors."[50]

That sort of slavering cupidity makes the baptist rationale look rather more attractive. And green energy does have an emotional and commonsense appeal that is hard to ignore. If we take off our green-tinted spectacles, however, it quickly becomes clear that 'green' is not an unalloyed synonym for all that is wonderful, and it is as well to remind ourselves of this. Green energy initiatives are fallible on at least two separate levels. One is in their sustainability as businesses, with or without various degrees of public support. The other is in their authenticity as truly green technologies – which is to say, carbon-free or at least carbon-lite – and hence in their capacity to deliver real, sum-total reductions in emissions. If they cannot deliver those, why are we spending all this money? Right now, green energy has another weakness, which is perhaps the one that should concern us most. Even if it were financially self-sustaining and totally emission-free, it could not be deployed on a sufficient scale to supply as much energy as we need, as reliably as we need it, for many years to come. This is not an anti-green book, but it is anti-waste – the waste of taxpayers' money. Green energy has a future but, like the Internet Bubble, it's all about timing.

The business sustainability of green energy companies matters precisely because so much is expected of them, and because the energy gap they are supposed to fill is so large. Most rely heavily on public subsidy in the shape of tax benefits, feed-in tariffs, free land, or state guarantees (which mean that the taxpayer picks up the tab if and when things go wrong). Although generally accepting that green energy is a 'good thing', consumers and taxpayers are for the most part unaware of the amount of money, their money, that goes into supporting current energy policies, often to relatively slight effect. Germany, for example, which is not a country famous for its sunshine, spent €35 billion in feed-in tariffs to expand solar to 0.6% of its electricity generation, according to a Ruhr University report published back in 2009.[51] More recently, in 2012, an energy professor at Berlin's Technical University estimated that renewable subsidies would cost German consumers more than €300 billion between now and 2030.[52]

In spite of these large sums, Germany is not yet attaining energy nirvana. Chancellor Angela Merkel wants renewables to account for 35% of German electricity by 2020 and 80% by 2050. The official Green lobby is more powerful here, and has more parliamentary representation than anywhere else in the world, which explains a lot about German energy policy. So it was for political rather than economic reasons that the government shut down the German nuclear industry after the Fukushima disaster in the spring of 2011. Its knee-jerk response prematurely expanded the hole that renewables must fill. Germany was already on an anti-nuclear course and its 17 nuclear power stations, producing about one quarter of its electricity, were due to close by 2036. Immediately after Fukushima, however, the government ordered the instant closure of eight of them and early decommissioning of the rest by 2022. With emission-free nuclear out of the picture, carbon-emitting gas will be used to take up some of the slack until, it is hoped, renewables can catch up. But the impact of an unpredictable and intermittent renewable energy supply is already being felt in faltering German grid stability, and industry has been complaining about damage due to sudden drops in voltage. Sales of industrial standalone emergency power systems have risen noticeably over the past year, according to reports in *Der Spiegel*.[53] In spite of turning its face against nuclear and coal generation, Germany is now obliged to keep the lights on by importing nuclear power from France and the Czech Republic and pressing on with the construction of nearly two dozen new coal-fired plants.

Germany is not alone in facing a huge bill to pay for its renewables aspirations. The UK, for example, wants renewables to account for 15% of all energy and 30% of electricity by 2020. Consultants at Ernst & Young have estimated that, for it to achieve its energy goals, £199 billion in new investment is needed by 2025, much of it to meet renewables and emissions targets.[54] The figure was revised down from an earlier assessment of £234 billion in the light of economic slow-down. Much of the opposition to renewables subsidy in the UK is directed towards wind. That is partly because, apart from the sometimes sunny south-west, Britain is not ideally suited to solar, and so there is not that much of it. But it is mainly because of rural opposition to the visual impact of wind turbines. Some have been trying to flag up the financial cost of wind energy. Ecologist John Etherington[55] estimates that over the 25-year life of a British onshore wind farm, it will benefit from subsidies via ROCs and tax breaks, but *not* including feed-in tariffs, of £3.5 million per installed megawatt. That compares with an initial turbine cost of about £1.25 million per megawatt. He pointed out that the effect of these two supports alone was to *double* the then wholesale price of electricity, adding another £60 per megawatt-hour to the £60 to £70 then prevailing.

If support is expensive for the consumer, it is positively mouth-watering for the investor – the bootlegger. Spanish researchers looked into the effects of generous energy subsidies (before they were frozen) and pointed out that, if the sponsors of a 100-kilowatt PV plant borrowed 70% of its cost, it would yield internal rates of return of up to 17%.[56]

Since the returns were guaranteed by the Spanish government, the risk was the same as buying a Spanish government bond. And yet a 30-year Spanish bond at that time was yielding closer to 5% a year. To give some idea of the scale of this generosity, the researchers noted that if you invested €100,000 at 17% a year for 25 years (the life of the guarantee), and reinvested the interest each year, you would end up with slightly more than €5 million.

So these subsidies help to inflate the Bubble. Since the business case is now dependent on political goodwill rather than the laws of economics, they also increase the risk of its bursting and the scale of the subsequent damage. Green energy projects typically need high levels of capital expenditure – which is to say it costs a lot of money to build them. If, as Etherington calculates, the installed *capacity* cost of UK onshore wind is £1.25 million per megawatt, but the turbine is only generating electricity 30% of the time, the cost per megawatt of actual *output* rises to £4.2 million. On the same basis, offshore wind costs nearly twice as much – £7 million to £8 million per megawatt of output.

The UK's Department of Energy and Climate Change (DECC) – in Britain, climate change now has its own government department – publishes estimates of the levelized cost per megawatt-hour over the lifetime of different technologies (see table opposite). The figures are broken down into component parts such as capital costs, operation and maintenance costs, fuel, and carbon costs. Bear in mind that estimating future costs of generation is more art than science, and often more wizardry than art. Champions of different technologies are not inclined to undersell themselves by overestimating their costs. And the figures may vary quite widely from country to country. For what it is worth, DECC's central capital cost estimates[57] show large solar to be the most expensive to build, at £143 per megawatt-hour. It is followed by offshore wind at £81 to £91, and onshore wind at £71 to £79. Next comes nuclear (£55), dedicated biomass (£38 to £52), coal (£22 to £26), and, cheapest of all to build, gas at a mere £9. Of course, the pecking order changes once lifetime fuel and maintenance costs are factored in, as you will observe in the table.

Levelised Cost Estimates for Projects Starting in 2012, 10% discount rate

SOURCE: UK Department of Energy and Climate Change

Central Levelised Costs, £/MWh	Gas - CCGT	Coal - ASC with FGD	Coal - IGCC	Nuclear - FOAK	Offshore R2	Offshore R3	Solar250 - 5000kW
Pre - Development Costs	0	0	1	5	4	6	-
Capital Costs	9	22	26	55	81	91	143
Fixed O&M	3	5	7	11	32	37	26
Variable O&M	0	1	1	3	1	-	-
Fuel Costs	48	28	30	5	-	-	-
Carbon Costs	19	45	56	-	-	-	-
CO2 transport and storage	-	-	-	-	-	-	-
Decomm and waste fund	-	-	-	2	-	-	-
Heat Revenues	-	-	-	-	-	-	-
TOTAL LEVELISED COST	80	102	122	81	118	134	169

Central Levelised Costs, £/MWh	Onshore >5 MW E&W*	Onshore >5 MW UK	Dedicated biomass >50MW	Dedicated biomass < 50MW	Co-firing Conventional	Biomass Conversion R3
Pre - Development Costs	2	2	1	2	-	2
Capital Costs	79	71	38	52	5	11
Fixed O&M	19	17	14	16	5	13
Variable O&M	3	3	4	5	1	1
Fuel Costs	-	-	65	41	81	83
Carbon Costs	-	-	-	-	-	-
CO2 transport and storage	-	-	-	-	-	-
Decomm and waste fund	-	-	-	-	-	-
Heat Revenues	-	-	-	-	-	-
TOTAL LEVELISED COST	80	102	122	81	118	134

* Estimates for onshore wind are shown using average load factors for UK and England and Wales 'E&W'

As we have seen, green energy projects are usually highly geared, which is to say that a high proportion of the money is borrowed, as debt, instead of coming out of the project sponsor's pockets as equity. That makes the venture all the more of a high-wire act. The exalted levels of debt ratchet up returns if things go well, but increase the scope for disaster if they do not. With banking and finance in its present enfeebled state, the scale of the debt involved means that, if a number of large projects all failed at more or less the same time, we could have another financial crisis on our hands.

And things do go wrong all the time. There are plenty of similarities between what is happening now in clean energy and what went on during the internet Bubble, but there is one very basic difference. The Internet exists in the cyberworld. It is all about software and programming, where the biggest cost, ultimately, is time – people's time or computing time. If it doesn't work, it's easy enough to pick up the pieces and start again. Energy projects are very much creatures of the physical world. They are essentially engineering projects, whether chemical, electrical, or mechanical, and when things go wrong here, they go badly and expensively wrong. Since the renewables industry is mostly still in its infancy, this technology risk remains very high. As noted earlier, expensive geothermal projects have ended with earthquakes or well blow-outs, and a number of wave projects have ended in tears. Wave Star was a Danish venture largely funded by Jørgen Mads Clausen, former CEO of Danfoss, the large thermal and energy controls engineering group. He reportedly sank some $20 million into the venture, which nonetheless ended in bankruptcy after internal squabbling over its future. Dexa Wave was another Danish wave energy initiative that filed for bankruptcy after its engineering challenges proved to be more long-lived than its funds.

We should not be surprised. The fact is that the odds in the wave business are heavily stacked against you. The technology risk is formidable and the capital expenditure demands are high. Maintenance and running costs are steep and the weather and the environment are out to get you. Much the same is true to a greater or lesser degree in other emerging green technologies, and the recent history of clean tech is littered with examples that simply didn't work, either technically or economically.

It was ever thus in the world of experimental engineering, so that is only to be expected. The only surprise is that we are more inclined to believe in the assured success of risky and expensive engineering projects if they happen to be green. And on top of the operational risk associated with any normal engineering project you must now add a thoroughly abnormal level of political risk.

Launching any new business is an act of courage, even of defiance, a declaration that it is possible to tame the forces of economics, the laws of supply and demand, the constraints of physics and chemistry and psychology, and to make them all work to your advantage. It is not an easy contest, but it is one where careful planning and good sense can give you a sporting chance of success. Politics, however, does not obey the laws of economic, physics, psychology, and the rest. It is irrational and subjective, at least when seen from business's point of view. A green venture that bases its viability on politically-motivated financial support is exposed to this irrationality, and to whatever the politicians decide to do next. The delicate internal financial architecture of most subsidized green projects is easily damaged. It does not take too much change in the level of subsidy for the sums to stop adding up and for the whole edifice to come tumbling down. Imagine a table, with one leg so much shorter than the other that it can only stand by being supported with a brick. The hope is that, over time, the leg will grow and eventually the table will stand unaided. If it does grow, the original brick may be replaced by a smaller one, and then a smaller one. The risk, completely beyond the table's control, is that someone whips away the brick altogether before the table can stand without it. As we shall see, politicians in some countries – Spain, for example, and Bulgaria – are already whipping away the brick.

There is nothing wrong with the principle of subsidy per se, within certain limits and as long as it does not last for too long. If subsidies encourage and bring on some useful industry to the point where, like our table, it can stand on its own four feet, there is a lot to be said for them. But they add risk all round. The risk for the recipient is that they may be taken away without warning. The risk for the taxpayer is that the beneficiary industry remains uncompetitive, year after year, becoming a semi-permanent drain on public funds or consumer bills.

To be fair, there is occasional evidence of subsidized green energy closing the grid parity gap. The costs and efficiencies of individual energy sources can vary substantially from country to country, region to region, plant to plant, so it is difficult to generalize. But there are a few pockets of solar that are either very close to grid parity or have already reached it. These are inevitably in places that get lots of sunshine, and may enjoy some other unusual characteristic that tilt the sums in their favour. One small but sunny 'country' claims to be the first on earth to be 100% solar powered. It is Tokelau, a group of small atolls in the South Pacific, governed by New Zealand. Known as the Union Islands until 1946, Tokelau previously depended on diesel generators, which provided electricity for only 16 hours a day and were both dirty and expensive. The New Zealand government advanced the $7 million required to build the project. Since the islands used to spend $825,000 a year on diesel imports, the scheme will pay for itself in eight-and-a-half years, all else being equal. The project's backers said that the islanders would now be able to spend more on social welfare.

Tokelau's not-so-near Pacific neighbour, Hawaii, also has a natural affinity with solar power. As an island state, Hawaii has depended on imported oil for more than 80% of its electricity generation and was, understandably, an early champion of renewable energy. The state introduced a 35% tax credit for rooftop solar hot water and PV electricity as far back as 1976. Much more recently, it introduced feed-in tariffs. These are especially generous in the case of commercial solar which, with the help of accelerated depreciation, can recoup its investment costs in as little as one year. Residential and commercial customers can also opt for 'net metering', which allows them to rewind their electricity meters as they generate power from their own roof-top solar installations. When they get their electricity bill from the utility, it shows the net amount – what they used minus what they generated. If they generate more than they use, they get paid. With the incentives, roof-top solar became somewhat cheaper than power from the grid, particularly as solar costs continued to fall, in Hawaii just as everywhere else in the world. Then the oil price started to climb.

As a result of the surge in the price of Hawaii's oil imports, by 2009 the price of local electricity was two-and-a-half times what consumers in the continental US were paying.[58] The falling price of solar electricity converged with the rising price of utility supply – grid parity! – in 2010, and the difference has been growing in solar's favour ever since. Today, solar power is a better deal for consumers even if they do not claim the tax credit or sell energy back to the grid. Residents can install roof-top solar and the savings will cover their outlay in 10 years, even without claiming any federal and state tax credits, claim the credits and the payback period shrinks to five years.

So it is not surprising that there has been a stampede into solar in Hawaii. Whereas the installed capacity was 25 megawatts in 2009, and 85 megawatts in 2011, it was 389 megawatts by the end of 2013.[59] Growing optimism has caused the state to accelerate its clean energy targets, from 20% of energy production by 2020 to 40% by 2030 and, eventually, to an ambitious 70% by 2030. Most of its existing renewable supply comes from wind and geothermal, both in abundant supply, and biomass (including bagasse and waste). It takes a lot of rooftop solar to make even a dent in the overall numbers and, so far, solar contributes a little more than 1% of the state's electricity. That is a surprisingly small proportion, given the persuasive economics, and Farrell (2012) cites a study claiming that Hawaii could satisfy 49% of its needs with rooftop solar alone. That may be true, but a number of obstacles have arisen. One is that in older homes installing solar means additional wiring and interconnection costs, which cannot be offset by incentives and which either extend the payback period or put people off altogether. Local authorities have been flooded with applications, leading to bottlenecks and delays, and the burgeoning market has attracted many fly-by-night contractors with limited experience. That adds to the need for inspections and creates more delays.

These problems are not intractable. What may be harder to resolve is the '15% rule'. This limits solar PV to a maximum of 15% of peak demand on distribution circuits – the local lines that connect homes and businesses to the main transmission network. It is a not uncommon safety regulation across US states, drawn up to cope with growing volumes of 'distributed generation' – the small-scale residential and business generation that feeds supply back into the grid. Existing networks were designed to deliver electricity from large power stations to customers, rather than to accommodate little gobs of electricity coming back in the opposite direction. Since the grid operators cannot control distributed generation, the rule is supposed to ensure local networks stay safe and stable. Once feed-in capacity hits 15% of the local total, either no more supply is accepted or would-be suppliers have to undertake a long and very expensive study (sometimes costing tens of thousands of dollars) of the project's acceptability for connection. That too puts off many people.

These difficulties will no doubt be dealt with in due course. There is some acceptance that 15% is an unreasonably conservative number, and the utilities may agree to it being raised. There is also a technological solution to the problem waiting somewhere in the wings. But the Hawaiian experience illustrates a number of pertinent points about green energy in general. One is that, given the right location, it can make an early escape from the hobbles of subsidy to compete on a level playing field with fossil fuels – as solar is doing in Spain and Greece. And yes, falling costs from economies of scale and technological advances will contribute to that. Yet even in sun-drenched Hawaii, making the break depended on the relative cost of fossil fuels. A steep rise in the oil price made the job a lot easier. Another point is that renewables make life difficult for the grid as it is presently constituted. And yes, the grid will have to accommodate them somehow, which brings us to the third point. It all takes time, and time, if we are going to keep the lights on in the near future, is not in abundant supply.

Our energy system – the legacy system, if you like – has been designed on a leviathan scale. It revolves around big baseload generation, pumping it out 24 hours a day, augmented by more flexible, perhaps smaller-scale plants when necessary. Wind and solar, by comparison, is will-o'-the-wisp stuff, sometimes on, sometimes off, not there when you need it, tapping you on the shoulder when you do not. As MacKay (2009) puts it, "most of the big renewables are not turn-off-and-onable"[60] and, if the grid is obliged to accept their supplies, as is often the case, that creates huge problems.

The most immediate problem is intermittency of generation. Consider the wind. The simple fact is that it does not blow either fast enough or slowly enough (or at all) to generate electricity for the majority of the time. Periods of little or no wind over Europe can last for days and occasionally weeks as high-pressure areas settle over the continent. The then energy minister Charles Hendry told the UK Parliament that the average load factor achieved by onshore wind farms in England in 2010 was only 20.8% compared with an average of 24.6% over the period 2000 to 2010. (The figures were readjusted later, but not substantially.) That was due to a lack of wind. Figures may differ around the world, and offshore wind typically has a higher load factor – it averaged 39% in the UK in 2013, its best year ever. But the intermittency occurs everywhere and it matters, because grid managers need to balance the supply of electricity with the demand at all times. Without getting too deeply involved in the principles of electricity, John Etherington sums up the problem thus, in *The Wind Farm Scam* (2009), his diatribe against the wind energy industry:

"Electricity must be used simultaneously as it is generated. If it is not, then the system becomes overloaded, [and] AC frequency and voltage rise until automatic frequency-sensitive controls ramp-down or trip-out some generating plant. Conversely, if demand outstrips generation, AC frequency falls, voltage falls and again automatic tripping occurs, this time switching out some consumers, if there is no spare generating capacity available to fill the gap. If these controls are inadequate, cascades of faults may cause failure of parts of, or the entire, grid system."

Grid management techniques and philosophy are evolving, and one of these days the big baseload model will be yesterday's story. Right now, however, it is still today's story and the system finds intermittent supply – and what is worse, unpredictably intermittent supply – painful to accommodate. What it really wants is turn-off-and-onable supply or, as they call it more soberly in the industry, "dispatchable" power, which is available when you want it. Solar is a bit more predictable than wind, but neither is anything close to dispatchable. Another measure of their reliability is their 'capacity credit', a concept allied to, but not the same as, load factor. Wind energy's capacity credit is the percentage of conventional generating capacity it could replace without making the system more unreliable. So if all wind turbines could generate at maximum capacity all the time, their capacity credit would be 100%. The UK National Grid, as cited by Etherington (2009), has estimated the capacity credit of UK wind at 20%. Etherington clearly does not believe this and points to the chief executive of EON.Netz, Germany's largest grid operator, who claimed that wind capacity credit in his country was 8% and would fall to 4% if Germany achieved its 2020 renewables targets. That is because the higher the installed capacity – ie the more the system depends on it – the lower the capacity credit.

In time, technology will make wind and solar power more dependable. There will be easier ways of storing it, so that it can be used when it is wanted and not just when it is generated. Smarter grids will be able to manage supply and demand and the general traffic around the system more intelligently. But for now, the low capacity credit of wind in particular means that, if the system is obliged to use it, it requires back-up. It needs reliable old hydro or a fossil-fired plant – something that fires up quickly, which means gas, but not nuclear – to step in when it cannot do the job. Now, that may not mean building new fossil-fired plants to begin with. But the more wind there is on the system, the lower the capacity credit, and eventually more back-up capacity will have to be built. That leads to the ridiculous situation in which, far from shutting down carbon-emitting generation, the onward march of intermittent green energy will actually create more of it. This extra capacity adds cost to the system, although it may not show up as directly attributable to wind. Helm (2012) reckons that the threshold beyond which these extra system costs start to kick in is when wind makes up about 20% of total capacity.[61] "The bad news," he notes, "is that the EU's targets will take all the main European countries over this threshold."

Something else that may add to the cost of wind energy is that wind farms are often in remote places, far away from where their electricity is needed. So new transmission lines have to be built. As former grid control engineer Derek Birkett points out in *When Will the Lights Go Out?* (2010)[62], the environmental impact of this can be as severe as that associated with wind farms themselves. Not only do the new generators have to be linked to the network but there is also a need for trunk reinforcement of the existing infrastructure. The cost of this is 'aggravated' by the "low utilization attending renewable generation".

This is not the place to get into the aesthetics of wind and solar farms, although the disfigurement of the landscape by windmills does rouse many a rural passion. Since higher is better for harvesting wind energy, the turbines are often found on the commanding heights, dominating the countryside around them. Campaigners against wind farms – and there are many, in many different countries – list the noise they make while spinning as one of their offensive attributes, and the fact that they kill birds and bats. The pro-wind lobby likes to respond that, as a dealer in avian death, the wind rotor pales in comparison with the domestic cat.

Whatever the effect on the bird population may be, as wind and solar energy become more embedded in the electricity system, they are playing havoc with the profitability of other generating sources, thanks to the 'merit order'. The controller of the grid is rather like the conductor of an orchestra. As demand rises and falls, he brings additional generation into play or shuts it down, with a sweep of his control-panel baton. There is no score, since the controller has to ad lib according to whatever consumers are drawing off the grid, but there are rules of engagement – called the merit order. Although it takes into account the flexibility and speed of start-up of different plants, the merit order essentially says that, all other things being equal, those with the lowest marginal cost of operation get asked to play first and are told to pipe down last. Those with the highest marginal cost are called upon last and silenced first. The marginal cost is principally the cost of the fuel needed to run the plant. So wind and solar, whose fuel costs are zero, are always at the head of the line. In the old days, the lowest-cost producers would often be the most heavily depreciated – ie oldest – fossil-fuel plants. The longer you operated them, the more money you made. But that has now been turned on its head, and the primacy of wind and solar – when available – has made other forms of generation less profitable. "On sunny and windy days, there is so much excess power that regional wholesale power prices have been driven to five-year lows," said a recent research note from Alexa Capital, a boutique investment bank specializing in energy.[63] 'This is great news for energy users that embrace technology to use electricity when surpluses are available, but bad news for power generators dependent on the old merit order.'

This merit order effect makes building new conventional generating plants – which, in current conditions, are likely to be gas-fired plants – a rather more questionable enterprise than it used to be, challenged by both lower prices and higher costs. The higher costs come from lower utilization and the fact that switching a plant on again and off again all the time pushes up unit production costs while creating more wear and tear than continuous operation – wear and tear that must be paid for in higher maintenance or shorter life, or both. A new gas plant will have to pay more for its fuel. It doesn't know in advance how often it will be called on to produce, so it cannot sign one of the take-or-pay gas contracts that offer the keenest prices. More wind calls for more back-up, even as it increases the costs and risks of providing it. If governments do nothing, not enough back-up will be built. So governments are doing something – many are now offering 'capacity payments' that guarantee a certain minimum income to new conventional generating capacity, even if it is not being used very much. That is another cost of having renewables on the system, and another one that, like the cost of additional transmission, does not show up as directly attributable to wind or solar.

Even Germany, which invented the green energy revolution almost single-handedly, is suffering from the crowding-out effect of unreliable renewables. The impact of so much subsidy-induced solar capacity now worries the head of the government-funded German Energy Agency, Stephan Kohler. "We have to make sure that operating power plants remain economically attractive," Kohler told *Der Spiegel*.[64] "Nowadays, solar systems are often in operation around noon, when there is high demand for power and the price was high in the past. As a result, conventional power plants can no longer make enough money, which is why existing plants are being shut down and no new ones are being built. Anyone who guarantees the security of supply in the future has to be paid for it, even if his power plant is only needed at certain times."

Kohler called for an end to the system whereby anyone in Germany who wants to install solar power, with the right to sell power to the grid, may do so. He described *Germany's Renewable Energy Act*, which created these rights and the subsidies that go with them, as "pure insanity".

If someone waved a wand and remedied all of wind and solar's shortcomings, they still could not provide all or even the bulk of our electricity. Even if they produced power economically without subsidy, day and night, without interruption, there is simply not enough space for the power farms that would be needed unless everyone wants to live underneath a wind turbine. David MacKay, who was once the chief scientific adviser at the UK's Department of Energy and Climate Change, calculated how much power could be produced if the windiest 10% of Britain (about half the size of Wales) was covered in wind turbines. He reckoned that was about as much density as would be physically or politically possible. His power estimate was the equivalent of 20 kilowatt-hours per person per day.[65] The UK average energy consumption per person per day is 125 kilowatt-hours, according to the *United Nations Development Programme Human Development Report* of 2007. The hardware required would be 50 times the entire wind turbine fleet of Denmark, seven times that in Germany, or twice as many turbines as there were then in the world (he published in 2009), but even then wind would contribute only 16% of total energy requirements.

We could push a lot of wind generation offshore, perhaps. But once you move turbines out into a maritime environment, the enterprise gets a whole lot harder and more expensive. Undersea cabling, fleets of support vessels, foul weather, and a hostile, wet, salty environment make installation and maintenance a costly challenge. Unlike its onshore brethren, it has not been getting any cheaper, either. The UK has more offshore wind capacity than the rest of the world put together. The UK Energy Research Centre calculated that, instead of falling, costs for UK offshore wind power doubled between 2004 and 2009, from roughly £1.5 million per megawatt to more than £3 million.[66] It did not foresee any meaningful reduction in the period to 2015. And here again, the number of turbines needed to make a substantial contribution is huge. MacKay imagines installing a strip of turbines four kilometres wide all the way around Britain's 3,000-kilometre coast. Such a strip would have an area of 13,000 square kilometres. It would deliver a mere 16 kilowatt-hours per person per day.

As we have already seen, politicians promoting the green cause like to claim that it creates jobs. That was as true in Spain during the recent boom years as anywhere else, as the government strained every sinew to encourage more solar and wind installations. Interestingly, researchers at Spain's King Juan Carlos University concluded that every new green megawatt installed led to the loss, on average, of 5.28 jobs, partly by raising energy costs and driving away electricity-intensive businesses such as iron and steel, chemicals, and cement producers.[67] They cite instances where rising Spanish electricity prices prompted such companies to relocate production to countries where energy was cheaper, including France, Poland, South Africa, and the US. The researchers observed that these job costs did not appear to be unique to the Spanish approach but were "largely inherent in schemes to promote renewable energy sources".

Where jobs are created, it may be a drain on, rather than a boost to, the economy. A newspaper recently revealed that, in 2012, UK wind turbine owners were paid a £1.2 billion subsidy via supplements on consumers' electricity bills. "They employed 12,000 people, to produce an effective £100,000 subsidy on each job," the report complained.[68]

If all of this takes some of the gleam out of green, then there are those sustainable energy sources that turn out not to be so sustainable at all, doing more harm than good, or requiring more energy to produce them than they can ever deliver as fuel.

The Innovation Center for Energy and Transportation, a think tank based in Beijing and Los Angeles, recently investigated the 'greenness' of China's electric vehicle (EV) industry. It discovered that, on average, electric cars and vans produced more greenhouse gas emissions per kilometre than their 'dirty' internal-combustion counterparts. "That's because EVs draw energy from electricity grids, some of which rely on coal – the dirtiest of fossil fuels – for up to 98% of generation," explained Lucia Green-Weiskel, a PhD student who worked on the project. "Only in regions where the grid depends on nuclear and hydro energy did the EV reduce its carbon footprint."[69] Although different regional grids have different fuel profiles, about 80% of all China's electricity is still generated by coal.[70]

The idea of "energy returned on energy invested" (EROEI) or "energy return on investment" (EROI) – take your pick – first surfaced in connection with oil in the early 1980s.[71] Two Cornell University ecologists, Charles Hall and Cutler Cleveland, noticed how US oil production had peaked in the 1970s and was now declining, even as drilling efforts continued to increase. They used 'yield per effort' concepts first deployed in fisheries studies to examine the energy relationship between drilling and actual oil finds, and published their original results in terms of barrels of oil recovered for every foot of well drilled. In the 1930s, US oil companies were producing 250 barrels for every foot drilled. By the late 1970s that had shrunk to between 10 and 15 barrels and by 1981 to a mere one-and-a-half. Later, they restated the results in terms that were easier to grasp – energy in vs. energy out. In 1930, they said, you got 100 barrels of oil back for every one barrel of oil invested in looking for it. In 1970, it was 25 to one. In the 1990s, it was between 11 and 18 to one and by 2005, when Hall presented the numbers at an oil and gas conference, he believed the figure was down to about three barrels for every one expended. Underpinning Hall's studies was the idea that as a fuel's EROI ratio approached 1:1, it was no longer useful to society.

If they hoped to make governments sit up and rethink our dependency on oil, Hall and Cleveland have been less than successful. But in the wider world of energy alternatives, the EROI principle has spawned a small industry, although some refer to it simply as a positive or negative 'energy balance'. Ethanol can have a significantly negative energy balance, particularly if it is made from maize or, as the Americans like to call it, corn. As far back as 2005, University of California Berkeley geoengineering professor Tad Patzek claimed that making corn-based ethanol used up more energy than the end product contained. He and his students considered all the energy inputs needed to extract alcohol from corn, as well as to produce the necessary fertilizers and insecticides, transport crops, and dispose of waste water. They estimated that the resulting ethanol contained 65% less usable energy than was consumed in the process of making it, pointing out that much of the energy input was fossil based. So anyone who thought that, by using ethanol, they were reducing fossil fuel emissions should think again. Patzek revised his figures in a later paper[72], but still maintained that the energy contained in ethanol was less, in varying degrees, than the energy required to produce it from corn, switchgrass, and wood biomass. And he and his co-writer said the same about biodiesel produced from soybean and sunflower oil.

This is a contentious area, not surprisingly, and others have disputed Patzek's figures. It is universally agreed, however, that corn-based ethanol is notably less energy efficient than ethanol made from Brazilian sugar cane. Add the impact of diverting food crops into fuel to the doubts about the energy balance from some sources and certain 'green' fuels appear less sustainable. The fact that the US government has been heavily subsidizing ethanol production in one way or another is less about 'greenness' and more about US ambitions for energy security. And it underscores the fact that simply describing something as 'green' should not be enough to give it a free pass, nor to blind us to its shortcomings.

CHAPTER 6:

CHEAP AND EASY GREEN

While we ponder the problem of how to feed our growing appetite for energy, we should not neglect one very important remedy – which is to use it more efficiently. Any energy that we can avoid using represents carbon dioxide we are not emitting and, frankly, that is the easiest, cheapest, and greenest form of energy there is. In fact, a whole 50% of the potential green energy market is in energy efficiency and it is an area to which investors are drawn because it is mostly made up of easy-to-pluck, low-hanging fruit.

There are various places where we can start looking for greater energy efficiency. One that most people know well is beneath the bonnet of the car. Energy efficiency makes a bit more sense to even the most self-absorbed of us each time the price of petrol goes up. The answer lies in technology or behaviour, or a mixture of both. More fuel-efficient engines are clearly desirable in this regard. Moving passengers and freight from motor vehicles to rail transport saves fuel, as does improved road transport logistics (more full truckloads, fewer empty ones). These are not new ideas.

In future, however, we may have to get used to thinking more about energy efficiency in our homes and offices. We will not be able to avoid it, because our electricity suppliers are changing the way they run their businesses, and because governments are getting tougher. What connects our electricity utilities and us, in more ways than one, is the grid, the network that takes in power from different generators and distributes it to consumers. The grid is getting more intelligent and, as it does so, it is enabling energy to be used more efficiently at both ends of the process. This idea of a modern, thinking grid trades under the name of the 'smart grid'.

The traditional, dumb grid transports electricity in one direction only, from big power plants to users, relying on manual intervention to juggle supply (which it can sort of control) against fluctuating demand (which, for the most part, it cannot). The people who run it do a remarkable job, but the fact remains that with the help of information technology the electricity industry can start to do some really clever things. It is much like what the phone companies did 20 years ago, using IT to make their own networks much more intelligent. The result was a whole directory of useful innovations from itemized billing and voicemail to having your calls follow you around, or setting a different ring tone for each member of the family. It was all about collecting data and then managing the information. The dizzying range of new functions provided by these IT advances went on to shape the mobile phone, and something similar (if less exciting) can be expected with our electricity experience. There is a downside, if you value your privacy, because by the time the smart grid is commonplace, 'they' will know even more about you and the way you live.

The smart grid is a loosely defined idea that embraces various technologies and benefits, and means slightly different things to different people. It includes communications overlay networks, data analytics, predictive controls, and regional grid integration. It may help to enable a shift from the hub-and-spoke architecture of today's generation and distribution, built around huge power plants, to more distributed microgeneration, but, and I emphasize the 'but', that will take time.

An important chunk of the smart grid has its roots in advances in meter technology. Automatic meter reading allows the utility to monitor your home or business electricity consumption without having to send someone round to read the meter. The next step in the evolution of the 'smart meter' is 'advanced meter infrastructure' (AMI), which allows the company to do a whole lot more. AMI provides two-way communications, allowing information and commands to be sent to the home. The utility pays different prices for electricity at different times of the day, depending on supply and demand. Until now it has averaged out the cost before charging home consumers a flat rate or, at best, operated a simple two-tariff system in which electricity is cheaper at night than during the day.

With AMI the company will be able to match the prices at which it buys and sells on a more individual basis. Because AMI lets it notify the user in real time of changing price bands, customers can choose to adapt their behaviour accordingly – by turning on the dishwasher at 10pm when power is cheap, for example, rather than at 6pm when it is not. 'Adapting your behaviour' may not necessarily mean rushing to check the meter every time you turn on the TV, but it may mean buying a smart dishwasher that knows the best time to turn itself on. The power company can use these time-of-use price signals to reduce demand during high-cost peak periods and encourage more of it during quieter times. That, in turn, will reduce the amount of spinning reserve (literally, turbines spinning but not generating, so that they can kick in quickly when needed) that has to be kept on standby. That uses less fuel, which is a good thing in itself, and saves the utility money, which should ultimately be reflected in consumer prices.

The load-smoothing possibilities move on a step further when the company offers residential users a lower tariff for 'interruptible' supply, something they have offered industry for years. This means that when demand is peaking and supply is stretched, they can in effect switch you off. It is called "load shedding". With AMI, load shedding becomes forensic, since the system can tell which appliances are on in the home, and then turn only some of them off – the air conditioning, perhaps. This 'demand response' feature goes down well in the US, where the grid struggles on hot afternoons when everyone is using their, yes, air conditioning. And it has other uses. Electropaulo, Brazil's biggest electricity distributor, plans to use this feature of the smart grid to combat electricity theft. The company, which has been losing nearly 4% of its power to thieves, will now be able to monitor consumption using wireless technology and, where it detects customers stealing electricity, will be able to cut them off remotely. Indeed, power companies everywhere like the idea of being able to disconnect non-payers remotely.

Because the system is now communicating with itself, it knows when and where there are problems. Having carried out the diagnosis, it may then be able to repair itself, or safeguard itself against failure. If a sub-station is blown out by lightning, for example, the problem area can be isolated, remotely, to allow the rest of the system to carry on functioning normally. In the US, Florida Power & Light (FPL) has installed 4.5 million smart meters across its jurisdiction as part of its Energy Smart Florida programme. If a power outage occurs, the company knows about it before the customer calls, because the meter fires off one last message before losing power. If neighbouring houses are sending similar communications, the utility knows it is an area problem rather than a one-off. If not, it may be an issue inside the home and the company can point this out when the customer calls. FPL says it resolved 32,000 customer problems remotely in a year, and reduced customer outages by 5.3 million minutes. That is a better life for customers and 5.3 million more minutes of revenue for FPL. The utility has also fitted early-warning maintenance monitors to its transformers, whose failure can result in serious damage to the system.

If the US interest in smart grids has concentrated on the demand response possibilities, in Europe, they have been more excited about the way in which they can help to balance the system, and accommodate intermittent renewables such as wind and solar. Unlike traditional power sources, which can be brought in and waved off at the behest of the controller, renewables are at the behest of the weather. If the weather is not being co-operative, and demand threatens to swamp supply, the smart grid can use price signals to reduce demand or it may begin load shedding unilaterally. The very idea has got some renewables champions thinking they can get rid of baseload supply altogether, inspiring articles with headlines like "Why baseload power is doomed".[73] This envisions a grid fed almost entirely by renewables, with a small element of flexible gas-fired capacity to cover the peaks. But it strikes me as overly optimistic to believe that this concept will easily be sold to the present generation of hard-nosed, hypercautious electricity engineers, for whom security of supply is the code by which they live.

Investment in smart grid technology more than doubled in the three years from 2010 to 2012, rising from $16.2 billion to $36.5 billion, according to London-based Memoori Business Intelligence.[74] Another UK energy research house, NRG Expert, brandishes even bigger figures. It believes that the international smart grid market will be worth $100 billion by 2020 and that full smart grid penetration will cost all of $2 trillion.[75] 'China's leaders view smart grid technology as "the next industrial revolution",' it claims, adding that the Chinese will have invested $90 billion in it by 2020. And it warns that electricity prices in Western countries will soar by 400% in the next 30 years "if electricity grids do not become smart grids".

Investors have been switching on to the possibilities they detect in this end of the green energy market, and in the US there has been an observable shift of venture capital and private-equity dollars out of solar and wind, and into smart grids.[76] Profits from renewables industries have dwindled, particularly in solar manufacturing as the effects of Chinese competition have been felt. There have been some spectacular US losses for investors – solar module manufacturer Solyndra, for example, absorbed more than $1.2 billion in venture capital funding before going into liquidation. Venture capitalists invested more than $490 million in panel maker MiaSole, which was subsequently sold to a Chinese company for about $30 million.

The fact that Silicon Valley venture capitalists are moving on is, in one sense, simply part of the natural order of things. They have always been the advance guard when it comes to technology and, as they move on to new pastures, big mainstream investors are moving into renewables, such as Warren Buffett with his MidAmerican Energy Holdings, and Blackstone Group. Venture capitalists are better at sniffing out new technologies than they are at building and running factories. Another area that is now attracting VC attention and money is energy storage, as they recognize that improved storage can make renewable power more grid-friendly, as well as raising efficiencies by smoothing production peaks.

The smart grid may deliver certain cost and convenience benefits to consumers (and that is undoubtedly how it will be sold to them), but its principal benefit for suppliers will be to slash the cost of meter reading. Some consultants question whether all the other supplier benefits will justify the added cost and complexity. Nonetheless, there will be some efficiencies at generating level as peaks are lowered. At the other end of the scale are the efficiencies that can be garnered by electricity users themselves. It is hard to find anything bad to say about using less energy. It is just good housekeeping, a no-brainer. We are conserving valuable resources or making them go further. Slower demand growth means less investment needed for capacity expansion. It is good for our bills and, if we import gas or oil, it is good for the country. No energy, as we have said, is the ultimate in green energy and, what is more, you do not have to subsidize it.

There are few areas of public, private, or commercial life where, if we tried, we could not use less energy. The European Commission has identified those sectors where it sees the greatest potential for savings, listing them as the building sector, with all its associated services, transport, industry, and even energy itself. Its list does not leave much out. But the EU has recognized that using less energy is integral to achieving its other ambitious targets, which is why it wants to reduce consumption by 20%, compared with projected levels, by 2020. If it could achieve that, it reckons it would save the equivalent of (pick one only) 2.6 billion barrels of oil, €193 billion, the output of 1,000 coal-fired power stations, or the GDP of Finland. Needless to say, it looks in no danger of hitting its target.[77]

There is general agreement that enough potential savings exist to meet the EU target 'cost-effectively' – in other words, in such a way that any investment is more than offset by the benefits. The scope for 'technical' improvement – cutting consumption and never mind the cost – is considerably higher. So if it is such a brilliant idea and it is do-able, why such slow progress? One reason is that energy efficiency is not very sexy politically. Schiellerup (2011) puts her finger on it: "Politicians get more applause for opening a new power plant than for implementing policies that make the need for a new plant redundant."[78]

Another reason is a variation on what they call the 'agency problem', whereby agents will often pursue their own, differing interests at the expense of those of their principals. Insulated houses and apartments require considerably less energy for heating or cooling but, in a rented property, whose interests does that serve? The landlord, as owner of the property, has to make the decision to insulate or not to insulate, and will have to pay for the investment. But he lacks the motivation because it is the tenant who pays the electricity bills, and who would reap the benefits of any lower outgoings. The tenant could pay for the insulation himself, but may not be planning to stick around long enough to get his money back. The landlord may find it difficult to persuade the tenant to make a financial contribution in return for the anticipated savings. Result? Nothing is done.

The state can take a hand. In the UK, a Green Deal scheme offers loans to households to pay for energy-saving measures such as insulation, double-glazing, or renewable self-generation. Unfortunately, this has not provoked a stampede of takers. One reason may be that people intuitively sensed what a national newspaper only recently quantified, which was that although the cost of installing energy-saving measures in an average house is a not inconsiderable £16,480, its estimated boost to the value of the home is, erm, £16,000.[79]

Another key to why we are not saving as much energy as we should lies in the way that energy-intensive new technologies have made us captives. Social anthropologist Harold Wilhite, a cofounder and first director of the European Council for an Energy Efficient Economy (ECEEE), describes how technologies like air-conditioning and refrigeration can displace ways of doing things that use much less, if any, energy.[80] Houses used to be built by craftsmen who understood how to make them as cool as possible, by means of location, orientation, and building materials. Today, those skills have been brushed aside, leaving us with hot-climate buildings that are uninhabitable without air conditioning.

What they call 'HVAC' – heating, ventilation, and air conditioning – systems are getting smarter, especially in the US, where local and federal authorities have been raising the efficiency bar for air conditioners.

But even with smart technology, America remains in air-conditioning's thrall. In Europe, which has not yet been colonized by the air conditioner, there is still time to arrest the process, Wilhite counsels, pointing out that "[n]atural cooling has been the norm for many generations of Portuguese, Italians, Spanish and French". Refrigeration has an even more aggressively colonizing effect, which now touches on almost every aspect of our food system. Tara Garnett of the Food Climate Research Network notes that, in 1970 in the UK, 40% of homes did not have a fridge and 97% of them did not have a freezer.[81] People stored food on shelves in relatively cool pantries and larders. But the advent of central heating and a desire to put the storage space to other uses spurred more refrigeration – which, of course, gobbles up more energy. Today, what Garnett calls "cold chain technology" is embedded in every stage of our food chain, affecting the kind of foods we like to eat, the way we shop, and the way we cook. There is no point declaring war on refrigeration, but some more creative thinking about how we can lessen its hold over us – and save valuable energy – may be in order.

There is scope for companies to score efficiencies by generating a portion of their own energy. In Germany, one in three companies is said to be doing this.[82] Volkswagen says it produces 60% of its own energy in its domestic market. In Brazil, where its locally made, Golf-like Gol has been the best-selling car since 1987, it plans to invest €640 million in PV, wind, and a second hydroelectric plant in the next five years. Self-generation may produce efficiencies from fewer transmission losses, and by lightening the load on the grid itself, although that is not always what motivates the companies concerned. Recent converts use a lot of wind and solar, and many will have taken this route to proclaim their green credentials as much as anything else. Ikea has announced that it will become energy independent by 2020, installing solar panels on all its stores and warehouses and investing in more wind farms, as part of a broader initiative to become greener.[83]

There is a bottom-line angle here, however, and the more cost-conscious will be looking for self-generation to pay for itself. Apart from the eventual cost benefits of growing your own, self-generators will get any available subsidies and premium feed-in tariffs, if they have surplus power to sell to the grid. They will also be able to sell their carbon credits. Some entrepreneurs see a business opportunity in advising companies on how to get into self-generation. But companies can be like people when it comes to saving energy, if they are not driven by ecological considerations. Since self-generation payback can stretch over a number of years, it can be an uphill struggle to convince executives who are inclined to think that, if funds are available, they would be better spent on a new production line.

Even so, companies can save energy like the rest of us by occupying more energy-efficient space and turning the lights off when they are not needed. Wilhite's work on energy efficiency argues that, at its best, it is an interaction between technology and behaviour, rather than just one or the other. Efficiencies can come from making a phone call instead of going round in person, or taking the bus instead of your car. But they can also come from having the right equipment, to control the room temperature, perhaps, or to turn off the lights automatically in unoccupied rooms. One of the most promising (and investible) advances in the 'having the right equipment' department is in light-emitting diodes, or LEDs.

The European Commission tells us that up to 50% of all electricity consumption in office buildings is in lighting.[84] The equivalent figure for residential buildings is a more modest 10% to 12%, but if less energy can be used in lighting, the potential for savings is considerable. And LEDs certainly use less energy. As part of its energy independence project, Ikea will change all lighting in its stores to LEDs, for example.

When we say that something produces 'more heat than light', the phrase must have been invented to describe the old incandescent bulb, which dates back to Thomas Edison and the first days of electric lighting. The underlying attraction of LEDs is that they produce a lot more light than heat. They are turning much more energy into what you really want instead of wasting it. In terms of watts per lumen (or power per unit of light), they leave Edison's bulbs standing. An LED lamp needs about 10 watts to give the same amount of light as a traditional 60-watt incandescent bulb. A compact fluorescent lamp (CFL) – those curly tube bulbs that so many of us are buying these days – is more efficient than an incandescent bulb but less so than the LED, requiring 15 watts. Give them all the same amount of power and an LED is roughly twice as bright as a CFL, and five times as bright as an incandescent bulb. This makes LEDs rather handy for street lighting or other applications where safety and security are important, like parking areas. The shares of Autev, a German maker of LED street lamps, enjoyed a burst of popularity when investors realized the potential for its products, although its recent delisting suggests that the company has since run into a technological or financial brick wall. But the street-lighting application flagged up some other virtues of LEDs, apart from energy efficiency and long life. Because the beams are more directional and less diffuse, they cause less light pollution; they are dimmable; and – a great blessing for some – because they emit so little heat, they do not attract insects.

Back in the home, the dimmable feature now means that, with wireless LED lighting, you can use your mobile phone or tablet to dim the lights or even to change their colour. Wherever you install them, the other towering advantage of LEDs is that, with no filament to be destroyed over time by heat, they have much longer lives, of perhaps 20 years or more. That is especially appealing in a street light, where changing bulbs requires a hoist or a ladder, but it is still pretty desirable anywhere else and makes them a very economic proposition, in spite of the high up-front cost of the bulbs themselves.

LEDs may also have a curious impact, via the insect world, on fruit and vegetable growing. A recent Dutch trial found that bumble bees can navigate better by LED lighting than by high-pressure sodium lamps when daylight is not available. The bees were active under LED lighting alone, but only became active under sodium lights when daylight was also present. Another disadvantage of sodium lamps is they generate heat that can be lethal to the bees. This discovery may seem to be of limited usefulness until you appreciate that lighting, and pollinating bees, are both important to year-round fresh produce growing in more northerly latitudes. Northern vegetable farmers may be heartened by trials in Finland, showing that aphids reproduce less under LEDs than they do under sodium.

LEDs have been around for a while as indicator lights and, more recently, as backlights for liquid crystal displays, typically in mobile phones and, latterly, televisions. Backlighting remains the primary application for LEDs and, although general lighting is catching up, it has not been catching up as fast as the industry would like. At an estimated \$17.7 billion[85], the global LED lighting market fell short of expectations in 2013, as consumers continued to resist high unit prices. But higher growth is predicted as prices come down. In 2009, a single 7.5-watt LED bulb cost about $120 in the US. Philips recently launched an equivalent that will sell for $10, although that still seems pricey to anyone who remembers paying $3 for an eight-pack of the now unavailable incandescent bulbs. Falling prices are a function of supply, and LED manufacturing has been looking a bit Bubbly of late. As ever, the Chinese have not been far from the scene. The Chinese authorities started to incentivize LED production in 2011, after which supply rose rapidly. Today, there are 4,000 companies in China making LEDs, although industry experts predicted that one in five would disappear in the course of 2013.[86] Even though prices have fallen, they are still too high for many Chinese households and, with the added aggravation of quality issues, domestic sales have been disappointing.

Nonetheless, although now is probably not the time to buy shares in a Chinese LED manufacturer, there is much bullishness about the industry's future. Consultants McKinsey & Co. reckon that LED's share of the general lighting market, which was a mere 9% in 2011, will be 45% by 2016 and nearly 70% by 2020.[88] In 2020, it expects the global general lighting market to be worth about €80 billion, so that is a share worth having, and it explains why so many people want a piece of it. A McKinsey survey of 4,000 executives showed that LED was regarded as the most promising technology in terms of commercial viability in 2020, followed by more efficient HVAC. The list included 13 kinds of clean technology, such as PV solar power, wind power, and electric vehicles. Carbon capture and storage was seen to be the least commercially promising, with cellulosic and algae biofuels coming second to last.

Investment has been pouring into the industry both from big established players such as GE, Philips, Osram, Nichia, and Cree, as well as from venture capitalists, who committed $800 million between 2008 and 2010, according to Cleantech Group. Some of the recipients include Bridgelux, Luminus Devices, Lattice Power, Lemnis Lighting, and Luxim, which has developed a non-LED 'light emitting plasma' technology, for very high brightness industrial applications.

The market is being driven, in one sense, by the inescapable logic of LED's superiority in the marketplace, even if this is taking time to filter through. But it is also very much driven by regulation. The energy-saving potential that can be released by the death of the incandescent bulb has not been lost on governments around the world. Many of them have now acted to phase out incandescent bulbs in one form or another. One of the first was Brazil. It has been followed by the EU, Japan, Korea, Russia, the US, and China, to name a few. Instead of a transition to the old technology's abolition, India and Venezuela have set up voluntary exchange programmes – new bulbs for old. It is this regulatory diktat that guarantees the future of LED lighting. CFL bulbs, with their irritating delay before full illumination, are filling the gap for now, because many consumers still will not swallow LED prices. But these are coming down and, even now, if you do the Maths, the annual running costs of LED are one tenth those of incandescent bulbs and about half those of CFLs.[89] McKinsey sums it up: 'The market is on a clear transition path from traditional lighting technologies to LED.' If we believe, as more and more people do, that energy saving is key to achieving our energy goals, this can only be good news. To repeat what I said at the start of this chapter, energy efficiency is the cheapest and easiest form of green energy.

CHAPTER 7:

EUROPE'S LOSING BET

O ver in the US of A they are pleased, perhaps even a little smug, about the way they appear to have solved their problems of energy supply and security almost overnight. Since the solution is based, to a large extent, on major discoveries of shale gas, it is not exactly carbon-free. But it is carbon-lite and, with national security anxieties now soothed, generators are already driving down US carbon emissions by switching from coal to cheap gas. The American situation looks pretty neat and organized compared with Europe, where different countries are pulling in different directions and emissions are going up, even though the EU has bet the farm on renewables. Some think that is already a losing bet.

When the European Council gathered in Brussels for one of its regular summit meetings in May 2013, there were two special items on the agenda: tax and energy. The European Commission had prepared an energy paper for the heads of government who make up the Council, and the contents did not give its readers any reason to feel at all smug.[90] As it pointed out in no uncertain terms, Europe is increasingly dependent on imports for its energy. It buys in the annual equivalent of €406 billion of oil, gas, and coal (3.2% of GDP 2012) and, in the case of oil and gas, these imports are expected to grow more than 80% by 2035. "Some member states rely on one single Russian supplier and often on one single supply route for 80% to 100% of their gas consumption," the paper noted. "This exposes them to the market power of their sole supplier whose price setting may not always follow a market rationale." Too true.

At the same time as Europe's imports are growing, it is having to compete more with others for these vital supplies, making them either more expensive or less available. That is because energy demand is on the increase everywhere, and nowhere more so than in Asia and the Middle East, where others may be quite happy to outbid Europe for resources. Because of the considerably higher prices offered by Japan and Korea for LNG, EU imports of LNG fell by 30% in 2012. Whereas Europe is a buyer in a seller's market, the US is moving from being a gas importer to a net exporter. The sheer abundance of US gas has widened the gap between industrial energy prices in America and in the EU. In 2012, the gas price for US industry was more than four times lower than it was in Europe, which makes European companies less competitive – painfully so, if they use a lot of energy. Between 2005 and 2012, gas prices fell 66% for US industry, but rose by 35% for European (OECD) industry, according to the IEA price index. They went up for US households, but only by 3%, and rose 45% for European households. Over the same period, real electricity prices for industry increased by 37% in Europe. In the US, they fell by 4%. European households saw their electricity bills go up by 22%, against only 8% in the US.[91]

The Commission paper also acknowledged, although not in so many words, that EU energy policy was failing at its very core, since coal usage was going up, not down. In Europe, gas prices are linked to oil prices and therefore rather high. As US generators dump coal and opt for gas on the grounds of cost, unwanted US coal is being offered for sale to Europe very cheaply. So European generators have been dumping gas and opting for coal. EU imports of coal were consequently up by 9% over the first 11 months of 2012, compared with the same period in 2011, although consumption was up by a more modest 2%. That masked some rather more pronounced individual increases in coal consumption – a doubling in Ireland, 38% in Portugal, 28% in Spain and the UK, and 16% in France. And as we know, more coal and less gas equals more carbon emissions.

A sound energy policy must deliver energy that is affordable, secure, and (ecologically and financially) sustainable. It seems that the EU is not doing very well on any of these fronts right now. Energy bills, imports, and carbon emissions are all going up. Its member states are spending varying sums of money to encourage green energy businesses that may or may not be sustainable. And yet investment in the EU's energy sector is now at historically low levels, even though nearly one fifth of its total coal capacity will be retired between now and 2050. That is equal to the total installed capacity in Poland. If you add the EU, Switzerland, and Norway together, known upcoming retirements of power plants are 70% greater than those in the previous five years. Not only is existing production going off the grid but planned new conventional production is also being cancelled, thanks to lower demand and lower returns as renewables force dispatchable power back down the grid queue. About 40GW of new gas and 25GW of new coal has been cancelled or postponed in the past three years, according to the Commission. That is the combined capacity of the Netherlands, Belgium, and Denmark.

One thing that vexes the Commission is the fact that some member states have surplus energy whereas others have shortages. Because of insufficient infrastructure connections with their neighbours, member states with shortages of energy may find themselves on 'energy islands'. Remedying this situation calls for investment of some €200 billion in transmission lines, interconnectors, and the like. But that is only the start. Then the EU president José Manuel Barroso told the assembled leaders himself that securing the EU's energy future calls for investment totalling €1.1 trillion by 2020. That breaks down into €500 billion on power generation, of which between €310 billion and €370 billion goes on renewables; and €600 billion on transmission and distribution.

A trillion euros? The IEA reckons that Europe needs to invest €3 trillion by 2030. Whose pocket does that come from? Well, the utilities, of course. Or maybe. Because Europe's power companies are mad as hell with Europe's politicians, and they are not trying to disguise it. A while back, they would not have cared because they were mostly state-owned, heavily regulated utilities that did what they were told. But today, many of them are privatized and deregulated, with shareholders to answer to. They are fed up with being taken for granted and being used as cash cows to be hit with windfall taxes whenever governments feel the pinch. Invest a trillion euros? How? They have been suffering from the recession along with the rest of us. Their shares are bombed out and their potential lenders and investors see the European energy market as a swamp that is best avoided. The utilities say that instead of being attracted to what is, or should be a stable industry with a steady long-term outlook, investors are actively deterred by its political and regulatory volatility.

Electricity generators, distributors, and traders across Europe and much of the Mediterranean are represented by Eurelectric. Its members are largely national trade bodies, so it is a trade association's trade association, which should give it a double taste for diplomatic and coded language. We can assume, therefore, that when Eurelectric starts to thump the table it really is reflecting the mood of the power utility industry at large. The organization has issued what can only be called an impassioned call for more realism, more EU alignment, and more innovation, with less politics and fewer targets.[92] When it asked 45 'energy leaders' if they thought this trillion-euro investment was feasible, 44 of them said "no". On average, they expected that only about half of that figure would be invested. Eurelectric points out that the assumptions underlying "these massive figures" were based on very different economic circumstances, and says that in the present climate the idea of investing those sums by 2020 is "simply unrealistic".

As we saw earlier, some countries are phasing out capacity and wondering how, and with what, to replace it. But in many of them, generators face the problem of reduced running hours for conventional plants. That is because this is being replaced in the merit order by renewable energy which, as Eurelectric sniffily puts it, "delivers energy when the sun shines or the wind blows, rather than in response to demand from customers". This phenomenon has made building new conventional capacity less attractive from a commercial point of view, even though it plays a vital role in providing back-up and security of supply.

Back-up capacity generally has to be gas or hydro, as the only currently viable resources that can be switched on at a moment's notice. So anything built specifically as back-up is going to be gas, given the very specific demands, not to mention scale, of hydro. The difference between what a generator pays for gas and the price it gets for selling electricity is called the "spark spread" (for coal it is the 'dark spread'). Deduct the cost of permits for carbon emissions and you get the 'clean spark spread'. Now, in the Central West European Market, which is more exposed than most to changes in gas prices, the clean spark spread is currently negative. In other words, the moment you switch on your gas turbine you are losing money because the gas costs more than the price of electricity – and that is before you have paid any other costs. Worse, it is not expected to return to positive territory until some time in 2018.[93] That is why countries such as France and Britain, which need new, reliable plants, are effectively bribing generators to oblige by introducing 'capacity remuneration mechanisms' that will pay new gas plants just for being there. Countries such as Italy, Spain, and Austria have more than enough capacity, and Eurelectric agrees with the Commission on the urgent need to connect such oversupplied regions with those less fortunate. Interconnection, however, takes a very long time to implement.

At this point, Eurelectric gets a bit more diplomatic. It knows better than to attack the sacred cow in public. So it is, it suggests, totally on board with the EU's desire for more renewables and "strongly supports the development of [renewables] technologies". But, it notes, money can only be spent once and when times are hard prioritisation of allocation is crucial. "The European electricity industry stresses the need to keep Europe's competitiveness in mind by prioritising economic efficiency as a key principle of the low-carbon transition," it says. It counsels a less national, more European approach, and "a greater reliance on markets and market-based principles to avoid stranded subsidies and stranded investments". That is a careful way of saying "Europe needs to compete, so let us not blow all our money on expensive, unreliable, uneconomic technologies that wreck the economics of those technologies we *can* rely on". I am sure Eurelectric will correct me if I am wrong.

Like me, Eurelectric is not opposed to renewables in principle. What it wants is an energy system that works, and one that we can all afford. But what it seems to want above all else is some consistency. That means consistency of policy across Europe, instead of 27 different national policies. Article 194 of the *Lisbon Treaty* guarantees member states the right to choose their energy sources and structure their supply, but Eurelectric would like them to be obliged to coordinate their policies. Consistency also means sticking to a particular policy once it is adopted

– no chopping and changing. That applies to renewables subsidies too. In an interview with the trade press, Eurelectric's head of energy policy and generation, Susanne Nies, cited Portugal's scrapping of its generous feed-in tariff for renewables, at the behest of the World Bank and others. That "devastated" Portugal's renewable energy sector, she said.[94] "We have a problem with politics," Nies continued, "with volatile, unpredictable policies upsetting the investment climate. Policymakers' job is to set a framework, not to take the entrepreneurial decisions. What happened to markets and competition?" What, indeed?

Her attitude reflects soundings taken among the wider electricity community. In 2012, the organization hosted a workshop to discuss the investment climate, attended by industry people, investors, and bankers. When surveyed, those present said that regulatory uncertainty and policy contradictions were the most important obstacles to investment. One banker pointed out that investors had opportunities other than the electricity sector and that, right now, they would rather avoid Europe and go elsewhere. Some big European utilities feel the same way, and are looking beyond the EU for business opportunities. Germany's E.ON recently spent €350 million on a 10% stake in MPX, a Brazilian power company with links to Chile. It suggested at the time that it might also be interested in Turkey and India. Spain's Iberdrola has already built up a position as one of the leading operators in Brazil's power sector.

If the companies themselves are feeling stressed, so is the investing and lending community. Banks have been in the eye of the Eurozone storm and are also having to worry about Basel III, which imposes stricter rules about how they allocate their capital. The rules herd them toward better-rated companies and shorter tenors. They also make project finance – long-term lending for projects like power-station construction – more difficult. At the same time, banks are having to pay more for their funds, so the lending environment is not pretty. Investors who have survived the grinding, volatile years since 2008, are still pretty jumpy and selective. Interestingly, the Eurelectric document singles out the Nordic region as being less exposed to investment difficulties than some others. It puts this down to "[a] greater regional approach, the structure of the generation portfolio, and more consistent policies…".[95]

For a view from the people who ultimately control the money – the investing and banking community – we can turn to Peter Atherton, former head of European utility sector research at the large US bank, Citigroup. Before leaving the bank, he gave a presentation to Future Energy Strategies, a London-based industry forum, entitled *Future of utility finance in the 2010s*.[96] He began by listing what he saw as the three historic goals of the sector, which were to deliver:

- affordable energy
- reliable energy
- safe energy

That was, presumably, before 'safe' got taken for granted, to be replaced by 'sustainable', But now, he said, it was also expected to:

- deliver climate change goals, for the whole economy
- drive an industrial renaissance
- create jobs and drive growth

Like Eurelectric, he too drew attention to the complicated matrix of policy instruments at EU and national level, asking how they all interacted, what impact they had on the value of new and existing assets, and what they actually left to the market to decide. Then he listed the assumptions that were driving EU energy policy:

- policy should be driven largely by the climate change objective
- the private sector can and will finance the €1 trillion-plus cost
- the consumer will be prepared to pay for the investment via tariffs
- Europe will not lose competitiveness, because the rest of the world will adopt similar policies
- because fossil fuels are increasingly scarce, their prices will keep rising;
- renewable technologies can be deployed on an industrial scale and therefore costs will fall
- the power system will keep working through this transition and beyond
- decarbonisation of the energy sector will be a huge net economic benefit, partly via first mover advantage.

The unspoken implication was that most or all of these assumptions were wrong. Then Atherton dropped his little bomb. "Economically," he said, "Europe is effectively taking out a massive €3 trillion futures contract on high fossil fuel prices." What he meant was that the way EU energy policy is structured and the huge investment it wants in renewables can only have a good outcome if fossil fuel prices keep rising. If they do not, renewables can never be competitive, and it will all have been a horrendously expensive mistake.

The capital markets do not really want to give European utilities the money to invest in the EU's plans, because they have been taking a dim view of them for some time. You can read their negative attitude in the share price charts. When economies are struggling, as they have been for some years, investors take refuge in defensive stocks, businesses whose profits do not go up dramatically in good times, but do not fall dramatically in bad times. That is usually because they sell goods or services that are not discretionary treats but are bought by people all the time. Drug companies and food retailers are classic examples. And so are utilities. In theory, power utility shares should have outperformed the market in general since the financial world had its own metaphorical 9/11, in the shape of Lehman Brothers's collapse. But in fact the shares of European utilities have *under-performed* the broader market, as well as under-performing the energy sectors in Asia-Pacific and Latin America. Shares of the biggest, such as E.ON, RWE, EDF, and Iberdrola, have been absolutely hammered.

Investors just do not like the risks or the prospects. When governments get into trouble they tax companies that cannot leave easily, and the fact that European utilities have been repeated victims of tax grabs – as the companies themselves have complained – has not escaped investors. Power companies are facing massive capital expenditure requirements and unrealistic targets for immature and difficult technologies. "Utilities are being asked to do too much too fast," Atherton said.

Meanwhile, back in the US, gas prices are down and the nation now has a century's worth of exploitable gas. Citigroup forecasts that the US will be self-sufficient in oil by 2017 and a substantial exporter by 2020. That will help to persuade offshored US industry to return to a reindustrialized America, create 3.5 million to 4.5 million new jobs by

2020, and reduce the US trade deficit by 56%.[97] Pause to catch breath. What if the US experience is duplicated, even if only to a limited extent, around the world, and gas prices fall back down to earth everywhere? We are only at the start of the shale gas revolution, after all. That would drive a stake through the heart of the EU's energy policy, because the economics of renewables rely heavily on higher fossil fuel prices to close the gap between them and grid parity. Even the European Parliament seems to be losing hope in renewables. Renewables badly need a high carbon price if they are to reach grid parity, because that will make using fossil fuels more costly. The carbon price needs to be about €30 a ton to do what it was designed for. In late 2014, European carbon permits were trading it at €6 a ton. That is largely because there are too many of them about. The year before, there was a motion before the European Parliament to cut the number of permits and effectively make it more expensive to pollute. It was narrowly voted down, suggesting that parliamentarians are more worried about keeping energy costs down during a slump than they are about combating emissions.

The EU set its energy targets largely between 2004 and 2007, and since then quite a lot has changed, Atherton noted. Such as: the financial crisis; the utility sector's abysmal stock market performance; disappointing performance data from European wind farms, Fukushima; falling energy demand, and the US energy revolution. "But EU ... targets and objectives have not changed one iota." Atherton also insisted that EU governments were in denial, before repeating his big point. "The EU is making a 20-year €3 trillion bet that fossil fuel prices will keep rising forever and ever and ever," he said. "The bet already looks a poor one to us."

CHAPTER 8:

BUBBLE TROUBLE

When I started this book I wanted to sound an early warning of what I feared would happen some time in the future. But even as I have been writing, the future has begun hurtling toward us in a disquieting way. In many countries, and across a number of different renewables structures and technologies, the Bubble simply goes on inflating, as sweeteners continue to be offered and as entrepreneurs take full advantage of them, borrowing large sums of money to do so. Yet the first signs of trouble are already appearing and in some places the Bubbles are already bursting. One reason is that some governments are waking up to reality and doing a U-turn on subsidies. You have just floated your boat and now someone is taking away the high tide they promised you. Another reason is that all this subsidy and government support has precipitated such a 'green rush', with such competitive overheating in parts of the renewables market, that they have reached premature saturation point. Although none of this is particularly surprising, it is all happening more quickly than I anticipated.

Let us remind ourselves of the basic premise behind green energy subsidy. Subsidies are a political tool, deployed to tilt the normal workings of the marketplace toward a desired end. In this case, the political goal is to reduce greenhouse emissions caused by burning the fossil fuels, coal, oil, and gas. Generous feed-in tariffs, loan guarantees, and tax credits are the carrot to attract new renewables generation, and mandatory limits on fossil fuel generation are the stick. But for every coal-fired power station that is closed down, a cleaner substitute must be ready to take its place. If the introduction of new but otherwise uneconomic renewables capacity cannot keep pace with the phasing out of old fossil fuel capacity, we have shot ourselves in the foot, expensively and rather painfully, because the lights will then start to go out.

Alistair Buchanan, chief executive of the UK energy regulator, Ofgem, opened a window on this possible future when he warned, early in

2013, that his country was heading for a 'horrendous' gas supply crunch that would force up electricity prices. (Independent research predicts that they will rise more than 50% by 2020.[98]) Over the next decade, Britain will lose about 20% of its generating capacity due to plant closures, because of old age or to meet EU environmental laws, or both. The old and dirty coal-fired power stations have each been allocated a specific number of valedictory generating hours, at the end of which their time is up. The problem is that they ate up those hours faster than expected to take advantage of cheap coal from the US and to avoid a new UK carbon tax coming into effect in 2013. As a result, capacity is being lost earlier than anticipated, with about 10% of the total being retired in 2013 alone. Wind and solar, which together contribute about 4% of the country's electricity, cannot begin to fill the gap. New nuclear will only come on line after 2020, if at all, so the slack will have to be taken up by new gas-fired power stations, which are quick and relatively cheap to build. Buchanan reckoned that by 2020, 60% to 70% of the nation's electricity will be coming from natural gas plants, compared with about 30% today.[99]

Two things really bothered him. One was that this new 'dash for gas' would almost certainly coincide with a global squeeze on the market for natural gas. That would force up prices and, since supplies from the UK's own North Sea gas fields are shrinking, the country would become hostage to foreign suppliers, such as – Buchanan did not name names – scary old Russia. He was also worried that the 'reserve margin' – the spare capacity available at times of peak demand – would fall from an already narrow 15% to 5% over the following three years, a margin he described as "uncomfortably tight". Although he would not go on record as saying he thought the lights would go out, it is self-evident that they could, given an abnormal spike in demand. This scenario is not unique to Britain, but applies to most other EU member states that are shutting down fossil fuel capacity and optimistically hoping to replace it with wind or solar.

What do the financial markets think about it all? As we have seen, they do not think much about the prospects for European electricity utilities. Until recently, however, they have been rather keen on green energy. Since the financial markets imploded in September 2008 and the global economy started to slow, there have not been many go-go sectors where you could invest your money in the hope of decent returns. In this negative environment, the guaranteed returns available in the green energy sector have allowed investors to walk on water. Not surprisingly, then, it has been one of the few sectors to have prospered since the crash. What is now Bloomberg New Energy Finance (BNEF) began tracking investment in 'clean energy' in 2004. Its definition of clean energy excludes large hydro-electric schemes and nuclear power. By its reckoning, the sector attracted total global investment of $54 billion in 2004. By 2011, that had leapt more than five times to a record $302 billion.[100]

Buried in the headline number for 2011 were a number of interesting features that told the story of what was really happening in the clean energy business.[101] One was that, unusually, a lot more money – about half of the total – went into solar than into wind. This was despite, or perhaps because of, the fact that prices of solar PV modules fell by nearly 50% in the course of the year. Wind accounted for one quarter of the year's investment total. It too was affected by lower turbine prices but also by a slow-down in China and various regulatory uncertainties in Europe. Another anomaly was that, for the first time since 2008, the US overtook China as the biggest investing nation. This was yet more evidence of the direct influence that sweeteners, or the lack of them, exert over the industry. The surge in US numbers reflected a stampede to get under the wire before the 2012 expiry of two important incentives: a federal loan guarantee scheme and a US Treasury grant programme.

Aspiring wind projects had another year to go before the expiry of a production tax credit for wind at the end of 2012. As a direct result, 2012 was a record year for the installation of new US wind-generating capacity. But overall, if 2011 was something of a binge year, 2012 was the year of the hangover. Worldwide investment fell by 11% to $269 billion, according to BNEF.[102] US investment fell by 32%. In two sunny Eurozone countries that had splurged on solar subsidies while getting their public finances into a mess, the drop was even more marked. Investment fell 51% in Italy and 68% (to a lowly $3bn) in Spain.

It seems that clean energy's infant growth spurt is over, at least in most of the developed world. Whereas the West pulled in its horns, the big spenders in 2012 were relative newcomers to renewable energy, such as South Africa and Saudi Arabia, both of whose drive into wind and solar was only just gathering momentum, and Japan, which introduced generous feed-in tariffs to encourage renewables following the Fukushima disaster. They are able to take advantage of falling prices in solar PV and onshore wind (although offshore wind prices have proved to be very sticky at high multiples of onshore wind).

Plunging prices are a feature of a manufacturing Bubble that, in the case of solar, has already burst. The technology was largely invented by the Americans and commercialized by the Europeans, most notably Germany. With a Green party powerful enough to move the political needle, and a historically motivated urge to be the best-behaved people in Europe rather than the worst, Germany is easily the most committed of all the major nations to the renewable energy ideal. It has been putting its money where its mouth is for more than 20 years, and has set itself formidable emissions and energy-efficiency goals. It intends to cut 1990 levels of greenhouse emissions by 40% by 2020 and, although the EU has not set any later targets, by 80% to 95% by 2050. Primary energy consumption must fall 20% below 2008 levels by 2020 and 50% below by 2050. Its generous subsidies fostered a large and successful manufacturing industry, particularly in solar, creating many thousands of European jobs, as the politicians had promised. But then the Chinese took an interest, and things began to change.

In 1997, when the Kyoto Protocol to the UN Framework Convention on Climate Change was being hammered out, China's carbon emissions were second only to the US, which it has since overtaken. As a developing nation, it was not obliged to accept any binding targets on future emissions. By the turn of the century, however, it had begun to worry about pollution and the sustainability of its coal-intensive energy mix. It identified solar PV as an area in which it could compete and started to encourage local manufacturers with cheap finance. Solar cells are not difficult to make and gradually the Chinese priced the Germans, the Americans, and just about everyone else out of the market. Today, China makes about 65% of the world's solar panels, according to the European Commission. The Commission takes a keen interest in this fact since, as it points out, Europe accounts for three-quarters of the global PV market.

This has led to some trade skirmishes, which have threatened to tip over into war. The US has accused China of illegal government subsidies and dumping prices, and has applied tariffs of up to 250% to Chinese solar imports. After investigating alleged Chinese subsidies, Brussels imposed punitive tariffs on Chinese panels. China's defence has been to attack, and it responded with its own investigations into US and EU exports of polysilicon, the key ingredient of solar panels. After the EU applied sanctions, it launched an investigation into European wine imports.

The good news was that prices fell dramatically which, along with rising efficiencies, has helped to bring these technologies closer to grid parity. As prices dropped, installed capacity shot up, from 9.4GW worldwide in 2007, to 23GW in 2009 and 101GW in 2012, according to the European Photovoltaic Industry Association. Germany, your sunshine destination, accounted for more than one third of it, with an extraordinary 36% share.[103] Italy had 18%. The rest of the world was a long way behind with Japan on 7%, Spain just under 7%, the US 6%, and China 4%. The bad news, for some, anyway, was that the competition started to put German and US manufacturers, who could not match Chinese prices, out of business. One of the highest-profile German casualties was Q-Cells. Founded in 1999 in Saxony-Anhalt, it produced its first PV cell, the part of the panel that actually converts light to electricity, in 2001. The Holy Grail of PV, like all energy production, is higher efficiency – maximising the proportion of solar input that is converted to power – and Q-Cells was a market leader in this department. It understood its vulnerability to price competition and set up a factory in Malaysia, which eventually accounted for nearly half of the company's production. But it was not enough to fend off the Chinese, whose cheap imports finally brought Q-Cells to its knees. It filed for bankruptcy in 2012 and the business was acquired by Korea's Hanwha the same year.

The very first major German casualty was Solon, one of Germany's biggest PV manufacturers, based in Berlin. After 15 years in business, it declared bankruptcy in December 2011. Most of its assets were eventually bought by solar manufacturer Microsol of the UAE. It had obviously been in trouble for some time, having secured €146 million in government loan guarantees two years earlier, after it was unable to borrow under standard market conditions. Solon was followed into bankruptcy only days later by Solar Millennium that, like Q-Cells, was a former stock market darling. It designed and built large-scale solar thermal power plants, including Europe's first parabolic trough complex, Andasol, in southern Spain, and it was planning another at Blythe in sunny California. The largest surviving German solar manufacturer, SolarWorld, lost increasing sums of money and has restructured its debt, taking in a new strategic investor – Qatar Solar Technologies. Not surprisingly, SolarWorld has led the Western solar industry's campaign for punitive EU tariffs on Chinese imports. Conergy, another former star in the German solar firmament, filed for insolvency before disposing of its manufacturing arms and selling its rump to a US private equity firm.

Perhaps the most totemic German casualty has been Siemens, the giant electrical conglomerate. Two decades of government support have built a strong pro-renewables lobby among German companies, most fervently of all within the German Engineering Federation. And the alpha male of German engineering is Siemens. So it was painfully symbolic when the company announced in October 2012 that it was pulling out of its loss-making solar power business. It blamed "the international pricing regime" for the losses behind the decision, and said its renewable energy business would now concentrate on wind and hydroelectric power. This was not the first time the company had reduced its energy options. A year earlier, it had pulled out of the nuclear energy business, despite having been involved in the construction of every German nuclear power plant. At the time, it had pointed the finger at the Fukushima disaster and the widespread German public and political fear of nuclear power.

Yet another big German name crashed out of solar recently. Bosch declared it would exit its loss-making solar panel production business by 2014, before selling it to none other than the recuperating SolarWorld. It has lost a staggering €2.4 billion since it got into the business in

2008.[104] When he announced the bad news, Bosch chairman Franz Fehrenbach said it was possibly the most painful experience of his entire career. It is significant that when it comes to choosing a solar panel brand for your solar farm these days, your decision turns less on technical performance – they are all within about 3% of one another, with a 3% margin of error – and a lot more on the warranty and the creditworthiness (and likelihood of survival) of the manufacturer.

Germany remains as committed as ever to renewable energy, particularly since it 'abolished' nuclear in the wake of Fukushima. Along with China and the US, it is one of the world's top three solar markets. But today the lion's share of solar panels sold in Germany are imported from China. David Buchan, a former energy editor of the *Financial Times*, summed up the situation in an instructive paper[105] on Germany's *Energiewende* – its much-trumpeted 'energy transformation'. "In effect," he observed, "German households have, through the renewable subsidies they pay, made the world a gift of solar technology which China has now been happy to exploit." The gift was not entirely metaphorical. Bizarrely, it turns out that one of the financial architects of the downfall of the German solar industry was the German government itself. More than €100 million from its development aid budget and other mechanisms designed to promote "global climate justice" has reportedly found its way to China[106], sometimes via roundabout routes. This has subsidized the drive of Chinese solar and wind companies into German markets. I do not know exactly what the Germans mean by "climate justice", but I suspect that is not it.

So, China one, Germany nil? Yes, and no. Because if the German solar industry is in trouble, so too is the Chinese. A combination of weak demand in Europe and lavish over-capacity (thanks largely to China) drove prices down by 75% in the four years to 2013. In 2007, annual solar panel capacity and demand were both below five gigawatts. By the end of 2012, demand had risen to 30 gigawatts, but potential Chinese supply alone was estimated to be 50 gigawatts or more. Widely thought to be selling panels at cost or below, and weighed down by debt, the biggest Chinese solar firms are staring bankruptcy in the face. The 10 largest have accumulated combined debts of nearly

$18 billion, according to Beijing's leading financial magazine, *Caijing*, which added that their accounts receivable position was worsening.[107] Boston-based GTM Research recently predicted that another 180 of the world's solar panel makers would collapse or be bought by 2015, and that more than 50 of them would be Chinese.

Since it was overzealous support that got the Chinese industry into this pickle, you might expect the authorities to whip away the drip feed. Not a bit of it. The response has been to lend them more money and to increase, rather dramatically, the targets for domestic solar penetration. Companies such as Suntech Power, which is the world's largest solar manufacturer, Yingli Green Energy, Trina Solar, and LDK Solar are all losing money. Yet they have still been able to increase their borrowings, often with the help of local governments loath to see regional jobs disappear. Claiming to combat air pollution, after 20 consecutive days of hazardous smog levels in Beijing, the Chinese government recently said it would raise its solar installation target for 2015 by two-thirds. Prices of Chinese solar corporate bonds, some of which are trading at extremely distressed levels, rose on the news. Yet this cannot solve the problem, and is merely booting it down the road to be dealt with at a later date, by which time the solution will have to be even more extreme. US and EU anti-dumping tariffs will not help. Increased domestic sales may boost volumes but, with prevailing price levels, they will do nothing for margins or profits until some of the competition is put out of its misery by acquisition or extinction. As I was completing this book, news broke that industry colossus Suntech Power had filed for bankruptcy. It seems that even the Chinese authorities have now seen enough blood. It is interesting to note that the industry's woes have not got in the way of personal enrichment for some fortunate solar and wind chief executives, who remain among China's richest men. Suntech was subsequently taken over by Shunfeng Photovoltaic International and, having been renamed Wuxi Suntech, appears to be coming back to life.

As the third of solar's three big markets, the US has not escaped torment. In August 2011, even before German solar companies started to file for bankruptcy, three US firms blazed that unhappy trail. The most prominent was Solyndra, an innovative Californian manufacturer. Instead of making the traditional flat solar panel, it produced tubular (or cylindrical, hence the company's name) solar receptors that, it claimed, produced significantly more electricity. The technology may have been unique, but it was not particularly cheap, making Solyndra vulnerable to a flood of incoming Chinese panels. Companies can come and go almost without notice in the teeming US marketplace, but Solyndra's passing caused a national stir. That was because it had been given $535 million in federal loan guarantees, enabling it to borrow on the cheap. Now that it was unable to repay its debts, the US taxpayer was going to have to pick up the tab, and the US taxpayer – or, at least, those opposed to green subsidies – was none too pleased. Abound Solar, a Colorado-based maker of thin film solar panels, was another failed recipient of loan guarantees, although it cost the taxpayer somewhat less. The company managed to get a $400 million loan guaranteed by the federal government, although by the time it filed for bankruptcy in July 2012, it had only drawn down $70 million.

Republican Party politicians pounced on the Solyndra case in particular as emblematic of President Obama's "wasteful" and "misguided" energy policy. They complained that three out of the first five US green energy companies to receive federal loan guarantees had gone bankrupt. (That was true but, as it turned out, they were the only three fatalities among the 33 companies eventually awarded guarantees.) But in October 2012, the conservative Heritage Foundation think tank published a list of 34 "faltering" companies that had received some form of federal support and were now either bankrupt (19 of them, accounting for nearly $15 billion in Department of Energy offers of support), heading for bankruptcy, or laying off workers. Republicans even introduced the *No More Solyndras* Act to review what remained of the loan guarantee programme. Although it was passed by the House of Representatives with its Republican majority, its promoters knew it could never get past the Democrat-controlled Senate. Nevertheless, they wanted to draw attention to the subject of green energy subsidy in what was a presidential election year.

The US federal government is not the only one handing out subsidies, which are also available at state and local level. In some states, such as New York, New Jersey, and California, state support can rival that from central government, and a combination of all levels of incentive can cover as much as 50% of a renewable energy project. At the federal level, however, they can be as intermittent as the energy they are trying to promote, depending on who is in the ascendant in the Congressional chambers – the generally pro-green Democrats or the generally anti-subsidy Republicans. One key driver of investment in wind energy has been a production tax credit that expired, not for the first time, at the end of 2012. Only a few days later it was extended for one more year as part of the fiscal cliff deal. Originally enacted in 1992, it has been extended by the US Congress five times since then but has also – perplexingly to anyone who does not understand the elephant-laboured-and-brought-forth-a-mouse nature of American politics – been allowed to lapse on four separate occasions. The installation of new US wind capacity has lurched up and down accordingly.

The unhappy course of the global renewables manufacturing industry is charted by the RENIXX World Renewable Energy Industrial Index, which tracks the share prices of the world's 30 largest companies in the renewables industry. It hit a peak of 1918.7 in December 2007. Five years later, by early December 2012, it had fallen by more than 90% to 159.4. By October 2014 it had recovered – somewhat – to aroundabout 360.

Wind turbine manufacturing around the world is not doing quite as badly as solar but it has troubles of its own, and for some of the same reasons. The Chinese piled into wind turbine manufacture later than they did into solar but they did so, once again, cheered on by the state. In 2005, there were 10 local and foreign wind turbine manufacturers present in the country and none of them were doing a huge amount of business. Then, in 2007, the government articulated its Green Leap Forward and promised meaningful expansion in solar, wind, and biofuels. This official blessing for wind energy ushered in incentives including cheap finance and cheap land. Earning official approval and earning money are overlapping and equally attractive concepts in China, and eager entrepreneurs tooled up, concentrating on the exploding domestic market. By 2010, the number of foreign wind players had not changed, but there were more than 70 Chinese manufacturers and the foreigners' share of the local wind business had reportedly fallen from 75% to 10.5%[108] – a smaller slice, but of a considerably bigger pie.

Chinese domestic wind capacity has risen so strongly and so fast that it recently overtook nuclear to become the country's third-largest energy source after coal and hydro, according to the Earth Policy Institute. There is often some distance between the picture painted by the numbers coming out of China and on-the-ground reality. Certainly, grid expansion has not been able to keep up with this wind farm explosion. Observers say that half of the wind capacity in north east China is not connected to the grid, and that elsewhere many turbines stand idle even when the wind is blowing.[109] (Grid connection has been a problem for solar as well.) The Global Wind Energy Council says, however, that China's total grid-connected wind capacity has now overtaken that of the US. The local market has slowed down even as it has become more competitive, and the official cheerleading has become less vociferous. Noises from the government now indicate that, although wind and solar should continue to grow, more reliance will be placed on hydroelectric and nuclear to meet China's renewables goals.

Nothing much seems to have changed on the ground. China accounted for 46% of the world market for new wind turbines in 2013, the World Wind Energy Association says. Nonetheless, the government has tightened up approval procedures to ease grid congestion, and China's turbine manufacturers have begun to look abroad more intently. Although the quality of their products does not match up to that of the best European and US manufacturers, their fierce domestic competition and lower cost base means they can be anything from 15% to 30% cheaper. This combination has not yet been compelling enough to duplicate solar PV's Mongol invasion of Western markets. However, the prices remain competitive, particularly when export customers can take advantage of cheap financing from Chinese banks, and the quality is improving. That is partly because Chinese wind farm developers are themselves demanding better quality.[110] The first Chinese turbines were only sold into Europe in February 2012, according to *The China Times*.[111] They were two Sinovel 3MW machines, bought by CRC Vindkraft, a Swedish wind power company.

Solar manufacturing is a volume business with all the price issues that flow from that, so its problems tend to look much the same wherever you go, be it Europe, the US, or Asia. In wind, where the hardware is a serious and expensive piece of engineering, with much longer lead times between order and delivery, the issues vary more from region to region. In the US, new wind yo-yos in sync with the on-again-off-again lure of federal supports. More recently, and perhaps even more disturbingly for its champions, US wind has become less rather than more competitive, as new shale gas production drives down the price of natural gas. Recent figures from the US Energy Information Administration[112] show the levelized cost (including capital and operating costs) of new wind power in the US is $96 per megawatt hour. Advanced clean-coal plants and nuclear check in at about $110, and advanced gas-plant generation can now cost as little as $63.

Back in Europe, egged on by government supports, installed continent-wide wind capacity broke the 100-gigawatt barrier in 2012, according to the European Wind Energy Association. At country level, China leads the world with about 91GW[113] installed, followed by the US with 61GW. Germany (34GW) is third in the country rankings, followed by Spain (23GW), India (20GW), and the UK (11GW). The highest growth rates are in eastern European states such as Romania (home to Fântânele-Cogealac, Europe's largest onshore wind farm, with 240 2.5MW GE turbines), Poland, Latvia, and Ukraine, although they are coming off a small installed base. More broadly, recent growth reflects orders placed before the Eurozone crisis intensified, raising political uncertainties and generally damaging confidence. In Spain, once a leading market, new wind construction is negligible now that subsidies have been stopped. Scaled-back subsidies elsewhere, regulatory uncertainties, and problems in connecting offshore installations to the grid have all contributed to a slump in European demand, compounding the fall-off in US and Chinese orders.

Market conditions, therefore, are not great and the strain is showing. The best-known name in wind turbines, and the most publicly distressed, is Vestas Wind Systems. It was founded in Denmark in 1945 as 'Vestjysk Stålteknik A/S' (West Jutland Steel Technology) and made household appliances, farm machinery, intercoolers, and hydraulic cranes before getting into wind turbines in 1979. In 2003, it merged with NEG Micon, another big Danish operator, to create by far the world's largest wind turbine manufacturer. It competes on quality rather than price, and has historically produced a very wide range of products. These cater for every conceivable need but are more costly to sustain than a narrower, more standardized catalogue. The quality of Vestas' engineering put it head and shoulders above the competition, enabling it to expand rapidly and internationally as wind energy caught on. At its peak it employed nearly 23,000 people and had manufacturing plants in a dozen countries, including the US, China, Spain, Australia, and Romania. The financial crisis of 2008 and the ensuing downturn blew it severely off course, and profits dropped by three-quarters in 2009. The company has looked vulnerable ever since, closing facilities, firing staff, selling operations, issuing profit warnings, replacing its chairman and chief financial officer, and, finally, sliding into escalating losses in 2011 and 2012. A turnaround programme managed to shrink losses in 2013, however, and the outlook is more positive.

Vestas has been the world's biggest turbine maker since 2000, when it dislodged NEG Micon, the company it merged with three years later. But in 2012 it was pushed into second place by GE Wind Energy of the US, according to BTM Consult, a Danish wind research house. Owned by General Electric, the American firm benefited from the scramble to build before the US tax credit supposedly lapsed. That same year, in a bruisingly symbolic lightening of its burden, Vestas sold its wind tower factory in Varde, Denmark to Titan Wind Systems, China's biggest maker of wind turbine towers.

In 2013, however, Vestas was back at number one in the BTM Consult table. As part of its revival, it agreed a 50/50 joint venture with Mitsubishi Heavy Industries to make offshore turbines – Vestas' weak spot. Mitsubishi, which builds ships, planes, and nuclear plants, is developing its first floating wind turbine, and has said it wants 10% of the global offshore wind market by 2017. Is Asia going to do to Western wind manufacturing what it has already done to Western solar firms?

Perhaps. As already noted, the engineering quality demands of wind turbines are in an altogether higher league than those of solar PV panels.

But Chinese wind manufacturers, having come from nowhere to the big league in next to no time, appear to be consolidating their position. In BTM Consult's 2010 global league, there were four of them in the top 10, including Sinovel, Goldwind, and United Power. Denmark's Vestas was, yet again, number one. In 2013, there were five Chinese manufacturers among the top 15, including Goldwind at number two, United Power and a now suffering Sinovel. GE was down at number five. Others in the top 10 included German market leader Enercon, Siemens, Spain's Gamesa, India's Suzlon, and Germany's Nordex.

A number of smaller Chinese manufacturers have disappeared, however, and they are not likely to be the last. There are more wounded, dying, or dead wind businesses in other parts of the world. Spanish manufacturer Gamesa was riding high while generous subsidies made. Spain, at one point, Europe's leading generator of wind energy. Now that they have been stopped, its fortunes have inverted, and it is losing money and cutting jobs. Suzlon, the Indian manufacturer, had the world's fifth largest market share in 2012 but it is suffering from the classic problems of hasty expansion. In 2007, it borrowed heavily to buy a controlling stake in REpower, a German turbine manufacturer. Deteriorating markets have made it harder to repay its debt and it recently defaulted on a repayment of more than $200 million in foreign convertible bonds. In the US, the ups and downs of federal support are taking their toll. Clipper Windpower, with a factory in Iowa (and $2.3 million in state support), builds some of the largest US turbines but has seen orders dwindle. After two changes of ownership, its new private equity owners have halved the workforce. Siemens, which has wind-related plants in Iowa, Kansas, and Florida, has shed 600 out of a total of 1,650 employees.

Germany has not escaped. German wind turbine manufacturer Fuhrländer filed for insolvency in September 2012, blaming delayed projects and postponed payments from customers. It had already tried cutting its workforce by 15% and refocusing on 'core' competencies, but was eventually undone by continuing pressure on turbine prices and soft demand. The fact that Europe's economic slowdown has tightened credit markets, making it hard to raise funds for development, did not help.

Although we hear a lot about it, offshore wind is still a tiny part of the wind universe. The Global Wind Energy Council tells us that at the end of 2013, the world could boast 318 gigawatts (or 318,105MW) of installed wind capacity. Offshore wind accounted for 7.046MW of that, or precisely 2.2%. Almost all of it can be found off the coasts of northern Europe, led by Britain, with 52% of the offshore total, and Denmark with 18%. Belgium comes next with 8%%, then Germany (7%), China (6%%) Netherlands (3.55%), and Sweden (3%). Finland, Ireland, Japan, and Korea have a limited amount of offshore capacity, and Norway and Portugal have just dipped their toes in the ocean. And that is it. There are no offshore wind farms to speak of in the US, although a number of projects are under development. In 2013, the sum of 1,631MW in new capacity was added to the world's total, nearly half of it in the UK. Europe sees offshore as essential to hitting its target of 20% of energy from renewables by 2020. It expects to have 40,000MW on stream by then, so it needs to get its skates on, particularly since offshore has been having an uncomfortable hiatus.

These are big, expensive projects, way beyond the reach of the entrepreneurial, fast-buck types who have traditionally developed smaller-scale onshore wind and solar. So the sponsors here tend to be large utilities, such as Dong, Germany's E.ON and RWE, SSE (formerly Scottish & Southern Energy), and Sweden's Vattenfall. Orders in the largest market, the UK, seemed to come to a halt in 2012. Some attributed this to the anti-wind remarks of a junior energy minister, later repudiated by his boss, and others to the stumbling progress of changes to the UK subsidy regime. In Germany, which has the next-largest installation programme, there are problems surrounding grid connection, or the lack of it. The country's plans include installing 25,000MW of offshore capacity by 2030, but in the first half of 2012 it managed to complete only 45MW. Dong stopped work on one wind farm, putting it back into its 'development pipeline', because no contract had been awarded for grid connection. RWE delayed a decision on another offshore project, because it was not clear who would pick up the tab if grid connection was delayed. "If the wind park can't work, it costs us €100 million a year," a RWE executive told reporters. Pending legislation that sought to fix the liability issue would charge developers and customers for installation risks. Developers, on the other hand, say that the risks are too high and cannot be insured, and warn that investors will stay away from German offshore wind.

No doubt the Germans will navigate their way around this hold-up, but it brings us back to the central point, which is that renewable energy is generally less than competitive with other fuels. So, if we really want it, someone has to put their hands into their pockets. And that someone, ultimately, is us. But how badly do we want it, and how much are we prepared to pay, in a stressed economic environment, knowing that renewables can only play a minority role for the foreseeable future?

Let us make the point again. I am all in favour of sustainable green energy. But 'sustainable' must be sustainable economically as well as sustainable in terms of resources. Subsidies can only make a positive contribution if they are in place for a strictly limited period, until the chosen energy source reaches grid parity. The arrival of the grid parity moment is accelerated if fossil fuel prices go up, as we saw in Hawaii's unusual circumstances and as anticipated by the European architects of the energy revolution. If fossil prices stay where they are or go down, as evidenced by the US shale gas transformation, grid parity is delayed – possibly indefinitely. Note that low US gas prices are already beginning to be imported into Europe. Centrica, the UK utility, recently signed a 20-year deal to buy liquefied natural gas from Cheniere Energy of the US. The annual volume is enough to supply 1.8 million UK homes and the price – here is the important part – is indexed to US gas prices. Low fossil-fuel prices defer the grid parity moment for renewables and, if subsidies are withdrawn before the wind or solar farm can pay its own way, the project will collapse. The equity invested in it will be lost and the debt raised to finance it, almost certainly secured on the project itself, will not be repaid.

Green energy subsidies are being cut across the developed world, including in Germany, France, Italy, the UK, and the US (where they peaked in 2009). This was reflected in a fall in total renewables investment in 2012. For some governments, reining in subsidies is an attempt to keep the financial returns for investors within reasonable bounds, and for others it is part of an austerity drive. Some are more hard-pressed than others. Egypt, which has been in sharp economic decline, spends about \$17 billion a year on fuel and energy subsidies and wants to cut that by 50% in the next five years.

Under normal circumstances, cuts only apply to new projects. Most solar and wind farms get long-term guarantees of premium feed-in tariffs, beneficial tax treatment, or other supports before going ahead. These typically last 20 or 25 years and are essential to the project's lifetime viability. But 'normal circumstances' no longer apply in many countries, and a lot can happen in 20 years. At the extreme, an incoming government could set its face so firmly against renewable subsidies that it introduced retroactive legislation to scrap them and to declare its predecessor's guarantees void. There would doubtless be a long queue at the door of the courts and howls from investors' embassies, but a determined government could get away with it, as long as it was prepared to endure a foreign investor boycott. And yet there are ways of finessing a commitment to pay over-generous or unaffordable incentives without completely tearing up the guarantees. We have already seen some of them, most notably in Spain, which overdosed expensively on subsidy-inspired renewables before it realized it was going broke.

To be fair to Spain, it does enjoy a lot of sun and wind, so the temptation to exploit them for 'cheap' and politically-correct energy must have been hard to resist. But in its infatuation with green energy – like all infatuations, the passion did not last – the government of Spain got at least two things wrong. Its subsidies were so generous that it got more wind and solar than it needed; and it chose, for political reasons, not to pass on the considerable increase in costs to the consumer. Since wind was a cheaper option than solar and requires less land, Spain cranked up solar subsidies just as others, such as Germany, were cutting back on them. Interestingly, Spain's largest energy group, Iberdrola, is a major investor in wind power but has mostly stayed away from solar. Its chairman, Sanchez Galán, has said that solar – and this is a Bubble alert – is "a financial product, not an energy solution".[114] What he means is that, rather than being positively useful in the real world, it principally serves the purposes of investors and lenders, just like the securitized mortgages that inflated the credit Bubble before 2007.

Spain has now stopped subsidizing any new capacity, but by the end of 2011, the last full year before subsidies ended, Spain's accumulating 'tariff deficit' – the difference between what consumers paid for electricity and what it actually cost to produce – had reached €24 billion. Spain's generating capacity (which includes a substantial quantity of new, subsidized gas generation) now stands at double the level of its peak demand. So although the good people of Spain may have been enjoying cheap green energy as consumers, as taxpayers they are paying through the nose for it – and more of it than they could ever need. This was a classic Bubble, in which injudicious incentives set off an avalanche of greed.

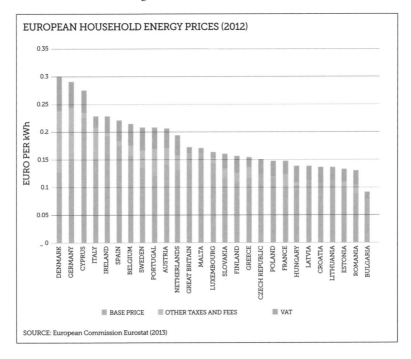

EUROPEAN HOUSEHOLD ENERGY PRICES (2012)

SOURCE: European Commission Eurostat (2013)

The Spanish government has clearly decided not only that it cannot afford to subsidize any more new capacity but also that it cannot really afford the existing stuff. So it has introduced tariff cuts and tax charges that will hurt existing producers and could put many of them out of business. It started back in 2008, when it placed annual limits on the amount of solar capacity that would qualify for subsidy and cut solar feed-in tariffs by 30%. A couple of years later it again cut solar tariffs and reduced wind tariffs by 35% for three years. It also decreed that there would be no solar support for a farm's first year of operation. Then it put the cat among the pigeons by *retroactively* cutting feed-in tariffs. Since the feed-in tariff law makes retrospective tariff cuts per se illegal, the government craftily reduced the number of hours for which PV electricity production qualified for the feed-in tariff premium, for a period of three years. More recently, it has changed the formula for payment to a fixed price based on the regulated tariff, and slapped a 7% tax on all electricity production. That too will have an adverse effect on green energy margins.

The investors concerned have, unsurprisingly, reached for their lawyers. A London law firm, Allen & Overy, has told Reuters that it is representing a group of international concentrated solar power (CSP) investors in relation to claims against Spain under the *Energy Charter Treaty*.[115] Signed by 51 states, the Treaty was the result of post-Cold War efforts to integrate the energy industries of the Soviet bloc with European and world markets. Today, it regulates international trade and investment in the energy sector. Since 2011, Allen & Overy has represented another group of disgruntled solar PV investors in similar claims against Spain.

The Czech Republic, another former green energy champion, has also had second thoughts about incentivizing renewables. Not only has it withdrawn support retroactively, but the new Czech President, Milos Zeman – a leftist, please note – has launched a bitter attack on the sector. He told parliament recently that solar subsidies had been the biggest theft in Czech history, stripping the economy of 200 billion crowns.[116] Previously, however, the country had embraced renewables with considerable zeal and, at one point, solar feed-in tariffs were 10 times the cost of power generated by CEZ, the national utility. In 2010, they were more than halved and various retroactive measures have been applied since then.

The Czech problem is that heavy industry, on which the economy still depends, is complaining loudly about the price of electricity. Much like the Spanish experience, this reduces Czech industrial competitiveness and deters foreign inward investment. Other measures affecting the viability of green Czech generation include the abolition of tax holidays, changes in depreciation, and an obligation to equip PV plants with facilities for remote power control. Causing the most indignation of all in the industry is a 'solar tax' of between 26% and 28%. The European Photovoltaic Association says that these combined moves have pushed the payback period for many PV installations 15 years beyond the 20-year lifetime guaranteed by the law.[117] PV power plants are operating at a loss, with negative cash flow and operational costs lower than revenues after the solar tax. The end result is that there have been several forced sales where entrepreneurs were unable to repay bank loans. That is what happens when Bubbles burst.

In Bulgaria, the Bubble burst with a particularly loud bang – as governments everywhere should take note. Here, mismanagement of energy subsidies and prices was not only financially costly but also led to blood-shed in the streets and the downfall of the ruling party. Bulgaria was the latest frontier in the green energy gold rush after it introduced generous feed-in tariffs for renewable energy investment. Soon, the country was getting more new green energy than it needed, and in 2012 it cut wind subsidies by 20% and solar subsidies by 50%. However, the costs already incurred were adding to the financial burden caused by higher gas prices and cost overruns at a new (and subsequently abandoned) nuclear power plant. The government introduced grid access payments for all renewable energy, ranging up to 39% of the applicable tariff. That is playing havoc with energy investors' carefully calculated sums. The government also banned PV installations on certain types of agricultural land, and gave the state regulator more flexibility to adjust tariffs at short notice. The unpredictability introduced by this last, critics say, means that from now on no bank will finance a renewables project in Bulgaria.

The Bulgarian government certainly upset foreign investors. When the subsidy cut was announced, China Ming Yang Wind Power Group was about to start construction of a 120MW wind farm in northern Bulgaria. It had already spent a sizeable chunk of the €150 million earmarked for the project but, since it had not yet connected to the grid, no longer qualified for the original tariffs on which it based its calculations. They were not happy bunnies. But the government also, fatally, upset its citizens. In mid-2012 it increased electricity prices by 14%. This did not immediately spark public unrest but steadily stoked resentment, which climaxed in the winter as electricity bills hit their seasonal peak. Bulgaria has the lowest bills in the EU, but it also has the lowest per capita income, and energy can account for a third to half of the poorest families' outgoings. The people hit the streets in February 2013 to demonstrate about the cost of electricity. After two weeks of increasingly violent protest, accompanied by calls to renationalize energy distributors, the Bulgarian government resigned.

The Estonian government has also been trying to reduce payments retroactively to renewable generators holding tariff guarantees. An increase in subsidies in 2007 attracted new foreign investment in renewables from companies including Finland's Fortum and Norway's Vardar. Then, in 2012, the government announced that it would cut subsidies by 15% to 20% to reduce the prices paid by electricity consumers. After some to-ing and fro-ing with investors, an agreement was reached whereby a ceiling on subsidized output was removed in return for the cuts. Then the government appeared to renege on the deal, reinstating the output cap while retaining the cuts.

The bootlegger brigade builds green power plants because they get a better return on their money than they would somewhere else, for taking the same or less risk. Utilities are energy professionals, not bootleggers, but they too need firm expectations of a respectable return before they will invest in renewable energy. If the required returns are only made possible by a subsidy of some kind, this needs to be guaranteed for the life of the project for it to proceed. If the subsidy is cut along the way, the returns will fall, quite possibly to zero, as we are seeing in places such as Spain and the Czech Republic. A cut in feed-in tariffs devastates the revenue side of the calculation. New taxes, levies, or grid access charges can do just as much damage by raising the costs of doing business. Of course, revenues and costs can be adversely affected by changes outside the regulatory sphere, and there is now some evidence that the wind industry may have miscalculated on both sides of the equation.

The Renewable Energy Foundation (REF) has published research showing that the economic life of onshore wind turbines is not the 20 to 25 years that the industry and governments have assumed, but 12 to 15 years. The reason for this is that the load factors – the amount of electricity actually generated as a percentage of the turbine's 24-hours-a-day capacity – decrease with age. In spite of its name, the UK-based REF is no apologist for green energy, but says it wishes to encourage informed debate on sustainable energy. The researcher in question is Dr Gordon Hughes, professor of economics at the University of Edinburgh and, until 2001, a senior adviser on energy and environmental policy at the World Bank. Studying the performance of wind farms in the UK and Denmark[118], he found that onshore load factors in the UK declined from a peak of about 24% at age one to 15% at age 10 and 11% at age 15. In Denmark, the decline was slower, from a peak of 22% to 18% at age 15. For offshore wind farms in Denmark the load factor fell from 39% at age zero to 15% at age ten. Hughes says the reasons cannot be fully assessed using the data, but adds that "outages due to mechanical breakdowns appear to be a contributory factor".

The declines mean that it will rarely be economic to operate wind farms for more than 12 to 15 years, after which they need to be replaced by new machines. As the REF puts it: 'The lifetime cost per unit (MWh) of electricity generated by wind power will be considerably higher than official estimates.' That will upset investors, because the cost of another machine will blow their 20- to 25-year projections to pieces. It will upset governments that have been relying on wind energy to help them meet carbon emission targets, because it suggests that the wind capacity and the capital investment required will be considerably more than they thought.

Hughes's data showed that larger wind farms had a systematically worse performance than smaller ones. He pointed out that since the average size of wind farms has increased, this has reinforced the deterioration in the performance of new wind farms. In the UK, he added, some investors were aware of these performance declines but continued to invest nonetheless. That suggested that the existing subsidy regime was extremely generous, and that there was room for further subsidy cuts to lower the cost to the consumer.

Cost to the consumer is at the very heart of this issue. It has already rocked the establishment in a poor country such as Bulgaria, but now it is starting to become a politically dangerous issue in wealthier nations. With most of its green energy jobs having marched off to the east, Germany's *Energiewende* may not have worked out exactly as the politicians envisaged. But it has also been very costly for German electricity consumers and will become even more so. They have the second-highest energy bills in Europe, paying about 25 euro cents per kilowatt hour.[119] Denmark, which uses more wind as a proportion of the total than any other country in the world, has the highest, at nearly 30 cents. Sweden pays 20 cents and France, which gets nearly 80% of its power from nuclear, pays about 14 cents. In Bulgaria, prices are the lowest in the EU, at just under 9 cents. German bills include a green levy that helps to fund feed-in tariffs. In 2013, the levy is due to rise by nearly 50% from 3.6 cents to 5.3 cents per kilowatt hour, costing the average three-person household another €175 a year. That has stirred the debate in Germany about the cost of green energy and who should pay for it. Electricity prices have now become a social issue, with the country's largest welfare organisation, the VdK, talking of increasing "energy poverty". In 2011, an estimated 200,000 welfare recipients had their electricity cut off because of unpaid bills, which the VdK described as a "glaring violation of basic social rights".

So on we go, ever deeper into Bubble territory. There are certain things that ring alarm bells; developments that have 'Bubble' written all over them. Joe Kennedy, father of US President JFK, famously said that when the shoeshine boy started giving you stock tips, it was time to get out of the stock market. It was a Bubble alert. One recent development that screamed "Bubble alert!" was summed up in an article in the *Financial Times*. The headline read, "Bankers turn sunshine into bonds".[120] I thought, "Uh-oh, here we go again." The story described how a company could lease solar panels to businesses and homeowners and then, instead of sitting back and collecting the lease payments for the next 20 years, package the rights to that income stream into bonds. This is the process known, with rather less affection these days, as securitization. The bonds would be sold to investors. Of course the solar companies would now have even more reason to lease as many solar panels as possible, emphasizing the tax breaks available (the whole idea was being floated in the US) to their potential customers.

Then there is the very entertaining game of 'leveraging the leverage'. Here is how it works: you create a solar farm costing €100 million, having split the costs into 20% equity (€20 million of your money) and 80% debt (€80 million of the bank's money). Once you have plugged in the solar farm and begun selling electricity to the grid, it enjoys a value uplift and is now worth, say, €120 million. You could sell it for €120 million and walk away. But instead you decide to build another solar farm.

You could repeat the financing process you went through to fund the first project, putting your hand in your pocket once again to come up with another €20 million in equity. But why would you do that if you did not have to? What you could do instead is to raise the money by refinancing the first solar farm. This is now generating electricity for guaranteed premium prices and is what bankers would call "derisked". So they might lend you money, secured against solar farm number one, on a loan-to-value basis of perhaps 85% or even 90%. And the value, remember, is no longer €100 million but now €120 million. If you borrowed 90% of that, you would have €108 million. That is more than enough for a 100% debt financing of solar farm number two, and you have not had to put in a penny of your own money. Brilliant. Well, brilliant until the politicians change their minds and something damages the revenue streams – like a retroactive tariff cut, or a grid access charge. Pretty soon, the company, which now owns two solar farms, is unable to service the debt repayments and the party is over – leaving twice as much debt. Leveraging the leverage was a great favourite in the days when collateralized debt obligations – another form of securitization – were all the rage. They helped to make the Bubble bigger and bigger, and ensured the damage was far worse when it popped. I have already seen a few examples of leveraging the leverage in the solar energy business.

This is how the early, happy days of securitization started, before the credit crunch. Now all we need is for leverage on leverage to spread, further securitization, then the slicing and dicing of the securities package into various rating bands and selling it all off to investors who have no idea what is really inside the package. The regulators have not learned much, and bankers will not be slow to catch on to this. It is easy, and it is exactly what they used to do. When feed-in tariff reductions kick in or extra charges are levied, the whole house of cards will start to crumble. Bubble alert.

CHAPTER 9:

THE RIGHT PRESCRIPTION

What should we do? Here we are, denouncing the waste of public money on renewable energy, but what is the alternative? How can we keep the lights on and the air clean enough to make it worth keeping the lights on? My starting point is not climate change itself but how governments have responded to the idea that our actions are affecting global temperatures. They are attempting to reduce carbon emissions through their energy policy choices. What I am questioning is whether they are the right policies. Will they meet their self-declared aim of lowering emissions while also delivering the three most important things we require from our energy: security, sustainability, and affordability? Sustainability and affordability go hand in hand, because our energy must be financially sustainable as well as sustainable in terms of resources. In an effort to kick-start renewable energy, many governments are channelling considerable sums of public money into the sector. Now, if there was no global warming and no scary rise in anthropogenic emissions, the case *against* spending a fortune of our money on renewable energy at all costs would be axiomatic, self-evident. What this book maintains is that our money is being misspent even if it is true that man-made emissions are rising and warming the planet.

It is always dangerous to be dogmatic about the future, but let us start with one safe prediction. In the years to come, the world will use a lot more energy than it does today. One reason is that there will simply be more of *us*. The planet's population continues to grow and is forecast to rise by a third from today's seven billion to more than nine billion by 2050.[121] Much of that growth will come not, as you might initially suspect, from China, which appears to be approaching peak population, but from India and Africa. Nigeria's population alone is expected to more than double from today's 170 million to 400 million.[122] Not only will there be more of us but also economic and social development mean that some of us will be consuming more energy than we used to. The developed world is trying to consume less energy, but developing nations are hitting the most energy-intensive phase in their history as they industrialize and as their citizens grow richer. As their disposable income rises and their lifestyles become more sophisticated, they travel more and buy more stuff, using up more energy directly and indirectly. The International Energy Agency (IEA) says that, even if governments keep all of their energy-saving promises, world consumption will still rise about 40% from 12 billion tons of oil equivalent (btoe) in 2009 to nearly 17 btoe in 2035.[123] Hewitt Crane and his colleagues measure energy in a different way, as we saw earlier, but they too anticipate a substantial rise in consumption, from today's three cubic miles of oil to six by 2050 – and that is only with 'markedly improved efficiency and a serious conservation effort'.[124]

So the challenge is not just to replace existing dirty capacity with clean generation but also to produce more energy than ever before. At the same time, this expansion must also offset all the bulk generating capacity, some of it nuclear, scheduled to shut down over the next few years because it is simply too old. The gap we need to fill is very large and, as I hope I have already made plain, the idea of filling all or even a significant part of it with renewables is fanciful. There will surely be a day when electricity generation and transport are 100% powered by clean fuels and a welcome day it will be, but it is further off than some would like and, when it dawns, neither wind nor solar is likely to be the number one player. The real issue is how to keep the lights on until then. We must have our energy drug one way or another, so what is the optimum mix required to feed our habit? What will be the most achievable and affordable energy policy?

Big oil companies like to think about the future, as well they might, and Royal Dutch/Shell has been better at it than most. To plan more effectively, the company pioneered what we now call "scenario planning". Until then, planning was mostly about making forecasts and extrapolating the present into the future. Scenario planning is a form of storytelling in which you imagine a future that you really want and then do what is necessary to make it happen. When Shell began scenario planning in the 1990s, it introduced the world to the concept of TINA – "there is no alternative". In the 2012 edition of its thought-provoking *Shell energy scenarios to 2050*[125], it introduced what it described as TINA's natural offspring, TANIA – "there are no ideal answers". Whenever we sit down to make choices about our energy future, TANIA should be looking over our shoulder.

So, in a less than ideal world, what is the best formula for our must-have drug? Clearly, we do not want to be over-dependent on any one energy source – that would threaten the 'security' element of our supply. And, as we have said again and again, our sources must be sustainable commercially as well as environmentally. As long as they can pay their own way within a reasonable time frame, renewables are welcome in the mix but at the core we need technologies that will deliver clean or, for the time being, *relatively* clean baseload power – in other words, a lot of it, all the time. We will come back to that. I am not opposed to the principle of subsidy, merely to the idea of prolonged and wasteful subsidy. So it is entirely positive that subsidy has helped solar to the point where it is now approaching grid parity in sunny countries such as Spain and Greece. I hope to see more of the same elsewhere. On the other hand, solar costs have been greatly helped by ruinous competition among panel manufacturers. It is quite possible – likely, even – that this phase will be followed by industry consolidation and some recovery in PV panel prices. It remains to be seen how resilient solar's increasingly competitive cost structure will be then.

If a limited period of subsidy has seen some solar and wind technology through to self-sufficiency, that is all well and good. Yet we cannot expect solar or wind to carry a major share of the generation burden. One reason is the fickleness of wind and sunshine, which means unreliable supply. That, in turn, means the more we rely on them, the more back-up we need from other, more dependable (mostly fossil-based, for now) sources. Another obstacle is the sheer cost and space required for wind and solar to make a meaningful contribution. Concentrating solar power (CSP) is the most suitable renewable source for utility-scale generation, although it is a long way from grid parity. Its flagship example is the Andasol complex in Spain's Andalusia, where the parabolic mirror troughs making up the first two units cover an area the size of 140 football fields and have a generating capacity of 100MW. Crane reckons that, for CSP to produce one cubic mile of oil – one third of today's total energy usage, or one sixth of it in 2050 – we would have to build 27 Andasols every week for the next 50 years, at a cost of $14 trillion.[126] The same capacity of rooftop PV would require systems to be installed on a quarter of a million roofs every day for the next 50 years. And the cost for that, even if current prices dropped five-fold, would still be $18 trillion.

Wind is going to be the exception rather than the rule. Not only does it suffer from the intermittency problem and the need for back-up but it has also attracted many enemies who resent the very sight of modern windmills. Offshore wind is more likely to be out of sight but the economics have a long way to go before they can live without subsidy. Siemens has begun making massive 6MW wind turbines that could be viable onshore without subsidy. Ironically, they are so large and so likely to attract hostility that operators are far more likely to install them out of sight offshore, where the economics are far less favourable. As lessons are learned, offshore costs will come down, but the market is 15 to 20 years behind the onshore environment. Meanwhile, back onshore, space remains an issue. Crane calculates that to get one cubic mile of oil from wind would require erecting 1,200 2MW turbines every week for the next 50 years. The results would cover an area three times the size of Spain.

Waste-to-energy can play a useful role. It is sustainable in the sense that we are not going to stop producing waste, but its future depends on how attractive it is to entrepreneurs. That, in turn, swings on how easy it is for them to raise equity for waste-to-energy projects, which can cost $40 million or more. It will also depend, to some extent, on input costs – in the US, waste-to-energy plants have to pay for their waste, whereas in the UK and parts of Europe they get paid to take it away. I have been looking at a waste-to-energy project in Dorset, assembled by a reputable operator, which will get only one third of its revenues from actually producing electricity. Another third will come from municipal 'gate fees' – being paid to take away the garbage they will burn – and the final third from green certificates. That is not exactly sustainable.

What happens, for example, when the municipality decides to charge for its waste? Plants that burn sawdust come in at a lower capital cost, but still need green certificates to be viable.

One of the Holier Grails must be geothermal. This is another example of a technology where subsidy should pay dividends in the end. In a perfect world, it can provide emission-free baseload power nonstop, 24 hours a day. The initial work is risky and expensive, with the precarious business of fracking, and the attendant risk of earthquakes, of the water not flowing as it should, and the whole enterprise nosediving as a result. So getting the first 10MW of a specific geothermal project up and running is hazardous and costly – the bill will be at least $300 million – and it will benefit from the encouragement of state support. But once you are there, it is very scalable. The next 10MW is pretty cheap. You just drill another hole and put some not very complicated, relatively inexpensive generating kit on top, which will cost you maybe $5 million. But it will be some time before geothermal becomes fully established. The marine technologies, loosely grouped under the headings of wave and tidal, also promise to provide 'green baseload', but their time horizon to grid parity lies even further out.

Some or, maybe, all of these technologies will have a part to play in our energy over time, but we need a couple of heavy lifters to produce the bulk of our baseload power over the medium term. Before we can determine what they should be, we have to remind ourselves exactly what it is that we want. Is it a) lower emissions, or b) more renewables? We need to get this straight in our heads because, even though these are two distinctly different goals, they are often confused with each other. Not only are they disparate ambitions but also – and here is the nub of it – one will cost us rather more than the other without necessarily being more effective.

That much was made plain in a 2012 report entitled *Powerful Targets*[127], put together by energy consultancy AF-Mercados UK, part of the Stockholm-based ÅF Group. It explored the relative cost of meeting decarbonization and renewables targets in the UK power industry between 2012 and 2050, via three different scenarios. Scenario one simply looked at the least cost way of meeting electricity demand, without any policy targets, but including a capacity margin at peak periods to ensure security of supply. It came up with a 5.8p/kWh unit cost of generation and a total cost of £780 billion. Scenario two was the same as scenario one, but included a requirement to meet targets for a 'significant' reduction in carbon dioxide emissions. Unit costs here were 7.2p/kWh, with total costs of £960 billion. Scenario three was the same as scenario two, but with the added requirement of meeting renewable electricity generation targets. Here, the unit costs reached 8.4p/kWh, and total costs were £1,100 billion.

In other words, meeting the UK's 2050 target for carbon emissions would cost 23% more than carrying on without a target. But adding separate renewables targets to that costs an extra 41%. Lots of numbers are bandied around the energy business, often by interested parties who distort them to flatter their own case. That phenomenon seems to have inspired this particular study that, the consultancy says, it prepared "because we feel that the debate on the costs of different energy futures often seems to be led by groups with vested interests". It insists its paper is not pro or anti any particular technology, but is an "independent perspective" on costs, based on data from reputable sources. The foreword to the report was written by Clare Spottiswoode. She is well-known in the UK as a high-flying economist who was director general of the then industry regulator Ofgas during the liberalisation of the gas industry in the 1990s.

"The model shows that the cost of having a renewables target over and above an emissions target alone is high," Spottiswoode wrote. "It is often not clear whether the aim of that policy is to reduce carbon dioxide emissions, or to deliver renewables for their own sake." If one was concerned about cost, then renewables had no part to play in reducing greenhouse gas emissions by 80% before 2050, she maintained. "Rather it is gas and nuclear alone that creates the least cost mix."

Given the economic impact, Spottiswoode said, it was important that the case for renewables was made independently and cogently. "There may be valid policy reasons to go for a costlier mix, but if this is the case, it needs to be articulated openly and honestly, giving stakeholders robust forecasts of the costs and benefits." She was writing about the

UK, but her observations could apply to energy policy in most, if not all, industrialized countries.

Now we are starting to close in on the limited choices available to us. Coal can give us non-stop bulk supply and it is cheap, but it is too dirty. We can clean it up by using carbon capture and storage (CCS), but right now that makes it prohibitively expensive. What else is there? As I said earlier, hydroelectric power comes close to being the perfect energy option. Although costly to build, hydro's levelized cost of operation – the all-in cost per unit of electricity over a plant's entire life – can be among the lowest of all. It is clean as a whistle, and can be turned on and off at a moment's notice, which makes it great for standby as well as baseload. Unlike most other sources, hydro plants can store electricity in bulk by means of pumped storage schemes. You do, of course, need big rivers. Brazil, blessed with big rivers, gets some 80% of its electricity from hydro plants, although this can have its downside in a prolonged drought.[128] China, which is starting to exhibit more concern about the environment, has been pushing ahead with more hydro construction, sometimes controversially. The big dams needed for meaningful hydro generation can only be built in certain topographies and can disturb the lives of many so, sadly, this is not an option available at any time, or any place.

So perhaps Spottiswoode put her finger on it when she said that only gas and nuclear can deliver affordably and in volume for the time being. Gas may not be emission-free but it is considerably cleaner than coal, so if it is fewer emissions we want rather than more renewables, gas is an obvious choice. Importantly, it is an affordable option. It has already become cheaper in the US and, although its price may not fall quite as dramatically elsewhere, the medium-term price outlook in other parts of the globe is down rather than up. New supplies of shale gas, together with copious quantities of LNG from Qatar and offshore Australia will, at worst, keep a lid on prices. Indeed, it would seem that the only direction for shale gas reserves is up – recent indications are that the UK has more reserves than everyone thought.[129] China may well be sitting on more shale gas than anybody else, and it plans to double its gas usage from the present 4% to 8% by 2016 – still a very small share by most standards. And then there is the unknown quantity that is the Arctic.

We may have underrated the extent to which gas will play a role in transport. In a special report on natural gas[130], *The Economist* notes that the US fleet of natural gas-powered vehicles doubled between 2003 and 2009, to 110,000 (since increased to 150,000, according to the Natural Gas Vehicle Coalition), although this is still a tiny fraction of the vehicle population. AT&T is buying 8,000 of them, which will make up the largest such fleet in the country, and 20% of local US buses already run on natural gas. The report quotes Michael Stoppard of research firm IHS CERA, who says that fully one third of the world's shipping fleet will be running on LNG by 2030. And if a growing global gas market makes gas cheaper than oil, that will further encourage the use of gas in transport.

Gas is certainly cleaner than oil, but it could be cleaner still. It is worth spending some public money to develop the 'clean tech' possibilities of natural gas. Energy research is a once-favoured child that has become neglected. Crane notes that in the OECD countries, the amount of money spent on developing new fuels and energy sources shot up after the oil embargoes of the 1970s.[131] But when oil prices fell, research and development spending fell too. The deregulation of the electricity industry also had the unplanned effect of reducing R&D expenditure, as utilities developed a new nose for profit and became keener to cut costs. Less R&D may translate into savings for utility customers in the short term, but in the long run it must hurt the whole energy sector. There is a role here for state encouragement of energy-related R&D via some form of subsidy, and it should foster more work on cleaner coal and gas – how best to extract the carbon dioxide from fossil fuels and use it profitably elsewhere. Cleaner gas is especially attractive because of the fuel's flexibility, which allows power generation to be switched on and off almost at will.

And then there is nuclear power, a technology that once seemed to hold the answer to all of our energy problems, only to become the pariah of the energy world. If the economics of energy markets is skewed by politicians, the effect reaches its extremes in renewables and nuclear. One has become a political darling, the other a leper. Unfortunately, the justifications for both stances are more emotional and sentimental – and so quickly exploited by a certain kind of politician – rather than factual. For the sake of our energy future, it is very important that we start to take some of the emotion out of the nuclear debate.

Nuclear is one of the cheapest, most powerful fuels, and safest – yes, I said safest – sources we have at our disposal. Nuclear power plants are expensive and time-consuming to build, but over their lifetime they can deliver some of the lowest levelized costs of operation. If the all-in cost of nuclear is competitive today, it becomes more so heading into the future, as carbon costs are expected to kick in, making coal and gas more expensive. A study for the UK government found that for projects starting in 2009, levelized costs of generation were approximately £80/MWh for gas, £95 for nuclear, £100 for coal, £150 for onshore wind, and more than £200 for offshore wind.[132] For projects starting in 2017, however, the pecking order changes thanks to the anticipated effects of carbon pricing. Nuclear comes in lowest of all at about £70/MWh, followed by gas on £95, coal £130, onshore wind £145, and offshore wind £180. These numbers refer specifically to the UK situation (the study does not even bother to look at solar, which is ironic since solar has now become the hottest renewables game in the UK for investors). However, the principle holds true in other countries that, in an age when money is tight, nuclear is actually a low-cost energy option.

If nuclear delivers a big bang for our buck, it also does it on a large scale. These are not little 10MW generators ticking over in the corner, but meaningful contributions to the nation's energy security. When Finland's new Olkiluoto 3 nuclear plant finally comes on stream, for example, it will have a nameplate capacity of 1,600MW. That is about 10% of Finland's entire generating capacity. And unlike wind and solar farms, nuclear plants are relatively compact and can deliver that kind of utility-scale power without taking up too much space.

Nuclear load factors – the amount of power plants actually produce as a proportion of their total 24-hours-a-day, seven-days-a-week capacity – are very high, certainly compared with wind and solar, and can run at 90% or more. This is real baseload generation, which can pump out electricity in large volumes, steadily, day after day.

It is also a secure option. The basic fuel – uranium or, as we shall discuss later, thorium – is not the exclusive preserve of some unstable or unfriendly government but is found in many parts of the world. We will always be able to get hold of it somehow. The fuel is inexpensive both in absolute terms and as a proportion of nuclear plant running costs, so even if its price were to rise substantially, this would have only a small impact on overall costs. Today, uranium sells for about $35 to $40 per pound. It hit an all-time high of $136/lb in 2007. Then, nuclear was seen as the next big thing, and demand was expected to rise. Supply prospects had been dimmed by floods at a big new Canadian uranium mine that had yet to open. At the same time, Russian supplies from its missiles-into-fuel-rods programme were expected to decline. We had seen this coming and, by buying uranium at the right time, made a lot of money. But the fact remains that, since uranium input is only 10% to 15% of a nuclear plant's running costs, even a large price rise is not the end of the world. That is not the case for oil- or gas-fired plants where fuel may account for 65% of the running cost.

Most importantly, if we really care about lowering carbon emissions, it is emission-free in its operation. The only substance that goes up a nuclear power station's chimney is water so, whatever anyone else may say, it is providing not merely baseload but *green* baseload, which is a rare and precious commodity. Over its entire lifecycle, a nuclear plant is responsible for some greenhouse gas emissions, if you take into account the raw materials used in construction of the plant, mining the fuel and so on. Measured in this more holistic way, nuclear is still cleaner than any other electricity fuel source except hydro and wind, according to a 2011 review[133] by the Intergovernmental Panel on Climate Change, the great Panjandrum of climate change science. The review found that the sum of all greenhouse gas emissions per kilowatt-hour associated with nuclear power was lower than that for biomass, solar thermal, geothermal, and solar PV.

If nuclear is so clean and cheap, why are we using less of it rather than more? The answer, of course, is because some of us think it is not safe. But how true is that? It is becoming increasingly clear that we have overstated the dangers of nuclear energy, both absolutely and relative to other fuels. That is undeniable if we count the number of people who die as a result of it. Both the European Commission and the Paul Scherrer Institute, which carries out energy-related ecological research, estimate that nuclear is the safest electricity generation technology, measured in deaths per gigawatt year.[134] Oil is the most deadly, followed by coal, largely reflecting fatalities in mining and exploration.

But what about nuclear radiation? Some people still insist on characterising nuclear energy as a death ray evil that will devour us if we give it half a chance. For some, such as Green politicians, this view is integral to their emotional orthodoxy and is unlikely to change. Others cleave to this notion simply because they are misinformed and because they have confused nuclear energy with the atom bomb. Some are not confused at all, but find it politically expedient to blur the boundary between nuclear disarmament and saying "no" to nuclear energy. This can have a political crowding-out effect. In Germany, however well-disposed mainstream politicians may feel toward renewables, their partiality is intensified by a need to keep the Greens on side (or at bay). The same is true, in a negative way, of their attitudes toward nuclear energy. This has led to the almost comical ideological contortion of ordering new (dirty) German coal-fired power stations so that (clean) nuclear plants can be shut down. One who argues strongly that we should moderate our fears of nuclear radiation is Oxford emeritus professor of physics Wade Allison. He points out that in the field of medicine, a lot of us now pay good money to receive much higher doses of radiation than we might ever be exposed to in a nuclear accident.

"How many radiation casualties have there been at Fukushima?" he asks on the website[135] for his book, *Radiation and Reason*. Answer: "None at all (and none are predicted in the next 50 years)." Allison recognizes nuclear angst as a muddled legacy of the Cold War, inseparable from images of ballooning mushroom clouds. However, whereas an exploding nuclear weapon produces a hugely destructive initial blast and fireball, civil nuclear technology can do no such thing. Yes, the impressive power of the technology comes from nuclear fission. "But this requires free neutrons that only exist within a working reactor," Allison notes.[136] "Otherwise there is only radioactive decay, and this radioactivity cannot spread by contagion like fire or disease. In a serious civil nuclear accident, the heat released by this decay can destroy a reactor. There are, however, almost no related deaths – none at Windscale (1957) or Three Mile Island (1979) and less than 50 at Chernobyl. An extraordinary safety record, thanks not to luck but to biology."

Allison's point, and one worth repeating, is that we live in a world where radiation is all around us. Natural radiation is not as weak as we sometimes think it is and artificial, man-made radiation is not necessarily as powerful. His message is let us be safe, but let us not be silly. Ultraviolet rays in sunlight can cause skin cancer but that does not mean we spend our holidays in dark caves. We use sun block and common sense. Nuclear radiation is not much different.

The worst civil nuclear disaster took place in 1986 when, thanks to operator error and design deficiencies, one of four reactors exploded at the Chernobyl power station in Ukraine. Although it caused anxieties around Europe and much nervous checking of radiation levels, fatalities were limited to 28 front-line firemen who died of 'Acute Radiation Syndrome' and 15 children who died of thyroid cancer, Allison claims. He says there is no firm evidence for any other related deaths and blames the "hurried evacuation of 116,000 local inhabitants" for "social and economic stress" that led to "depression, suicides, alcoholism, family break-up and broken livelihoods".

Then came Fukushima, in 2011. It may strike the disinterested observer that the main nuclear safety lesson of Fukushima was 'don't build your nuclear plant in an earthquake zone'. It is a testament to the relative safety of nuclear that the catastrophe has not resulted in any radiation-related deaths. But those who had always opposed the technology chose to represent the disaster as yet more proof of its dangers, and a general flap ensued. Germany, Belgium, and Switzerland promptly swore off nuclear, like lapsed drinkers promising to be good from now on. Italy, having sworn off nuclear in 1987 and then recanted, clamoured for another referendum and swore off it again. Plans for a new Italian nuclear plant had to be scrapped. It is exactly that kind of political risk which has put the private sector off new nuclear projects in the absence of cast-iron guarantees and financial support.

Whatever lessons were learned from Chernobyl, a measured response was not one of them. In 2011, the Japanese government set the maximum permitted radiation level for beef even lower than the Norwegians had done for reindeer meat in 1986. Either no one told them or they ignored the fact that the Norwegian government had realized the error of its ways only a few months later and raised the limit tenfold. As Allison puts it, after Fukushima you would have had to eat 1,000 kilograms of Japanese beef in four months to get the same radiation dose as that received in a couple of hours during a regular medical scan. He calls for the risks of ionizing radiation to be balanced against the benefits in the light of experience. And one of the benefits is as a safe, cheap alternative to fossil fuels, generating electricity to be used in the traditional way or as clean fuel for a new generation of electric vehicles. "New realistic safety regulations should bring large cost savings to any nuclear programme," he says. "While no corners should be cut in respect of the control of reactor stability and its heat output, with fresh justifiable standards many costs of nuclear power could be reduced dramatically and safely."

Even if we get less spooked by nuclear radiation than we used to, radioactive waste disposal remains one of the sticking points in the nuclear debate. Another is that if civil nuclear by-products, which is to say plutonium, get into the wrong hands, they can be used to make some very nasty weapons. As it stands, the need to store nuclear waste safely and for a very long time is simply one of the technology's less attractive features (TANIA, remember?), although one which is manageable. A change in fuel from uranium to thorium, however, would go a long way to address both the disposal and the weapons issues.

Thorium is not new to us. It was a contender back in the early days of civil nuclear, and the Oak Ridge National Laboratory in Tennessee devoted itself to developing a liquid fluoride thorium reactor (LFTR) during the 1960s. It lost the competition against solid core uranium reactors, partly because development of the latter was more advanced but also, importantly, because the US military establishment coveted uranium's ability to bring forth plutonium. Thorium does not do this, at least not without great difficulty. And that is one of the features that makes thorium so attractive today. Another is that it is about four times more abundant than uranium. Richard Martin is illuminating on both the benefits and the history of the metal in his book, *SuperFuel*. In it, he argues that if you were going to design a new reactor from scratch today, it would look pretty much like an LFTR. Some of its virtues derive from thorium itself and others from the reactor design, particularly its liquid, as opposed to solid, core.

"The only truly inherently safe reactor is a liquid-core reactor," Martin declares.[137] Without getting blinded by the science, that is partly because it operates at normal atmospheric pressure. Solid core, water-cooled reactors generally operate under extremely high pressure, at 60 or 70 'atmospheres', like bombs waiting to explode. You are continually trying to keep them from going out of control. They need an external power source to cool the reaction chamber and to shut down the plant if things start going wrong. Fukushima's problems started when the tsunami cut power to the plant and simultaneously swamped the back-up generators. Safety in a liquid (actually a molten salt) core reactor works the other way around, on the 'dead man's handle' principle. You are continually trying to keep the liquid reactive. It cannot melt down because the fuel is already molten. If things start to go wrong, physics takes over and the system shuts itself down. When power is cut off or the liquid gets too hot, a frozen plug at the bottom of the reactor melts, and the liquid drains away of its own accord. They call it "walk-away safe".

Because these reactors do not require large safety containments, they are smaller and cheaper. They are also much, much more efficient. Solid fuel rods suffer damage and create unpleasant by-products, so they must be replaced regularly after only a fraction of their energy has been used up. LFTRs do not have that problem and use up nearly all the nuclear energy, which makes the thorium up to 200 times more efficient. That means they leave much less waste. Best of all, perhaps, the LFTR can 'burn' the unused energy in conventional spent fuel rods – aka nuclear waste – which goes a large way toward solving our waste disposal problem. What's not to like?

Thorium has its enemies, in the shape of most of the existing nuclear industry and their friends in politics and the military. They have 'uranium' tattooed on their foreheads. It is what they know and how they make their money, so why would they want to change? That is especially true in the US. But as the thorium lobby likes to say: you would not want a 1950s car or a 1950s computer, so why would you want a 1950s nuclear reactor? Countries that are less financially, scientifically, and emotionally invested in uranium are more prepared to give it a whirl. China, keenly aware of a looming energy shortage, is leading the field. It reportedly already has 140 PhD scientists working on a thorium project[138]. The government of India, which has large thorium reserves but lacks indigenous uranium, says that thorium plays a "pivotal role" in its nuclear power programme. Norway's suitably named Thor Energy is involved in a test programme to see if they can use thorium in Oslo's conventional Halden research reactor. Some think that Japan has a major thorium programme up its sleeve.[139]

That is probably enough about thorium, which still has to prove itself on a commercial scale and is not going to replace uranium in a hurry. But it tells us that nuclear has a positive future. It is important to recognize that the nuclear industry is at last being revitalized and energized with new thought. It is going somewhere, and not just up a dead-end street. After Chernobyl, a generation of young scientists largely avoided the industry but bright students are now embracing it once again, bringing new ideas with them.

In its current state of evolution, nuclear is not without risks and costs. But (TANIA!) nor is any other technology. Coal is either cheap and dirty or cleaner but costly. Gas is cheap and less dirty but we cannot be sure what it will cost in future and, for those of us who do not sit on top of lots of it, it raises security of supply issues. Wind and solar are unreliable and, in many cases, expensive. The perception of nuclear as so risky that it does not merit inclusion in our energy formula is simply wrong, and it is heartening to see how even some card-carrying Greens and professional earth lovers are coming around to this view. One of the more notable is James Lovelock, environmentalist icon and author of the Gaia (as in Greek earth goddess) theory that unites the living and non-living elements of the earth into a single self-regulated organism. Fiercely protective of his Gaia-earth, he has been outspoken on the issue of carbon emissions and global warming. In 2004, however, he scandalized environmentalists with a newspaper article championing nuclear energy as the only logical replacement for dirty fuels like coal and oil.[140] Lovelock said – and he was correct – there was no chance that renewables such as the wind and the tides could produce enough energy soon enough to combat global warming.

"If we had 50 years or more we might make these our main sources," Lovelock wrote. "But we do not have 50 years." He echoed the arguments of Wade Allison, adding that the fears expressed by some environmentalists were unjustified, and that nuclear energy had proved to be the safest of all energy sources. "We must stop fretting over the minute statistical risks of cancer from chemicals or radiation. Even if they were right about its dangers, and they are not, its worldwide use … would pose an insignificant threat compared with … lethal heat waves and sea levels rising to drown every coastal city of the world."

Lovelock was motivated by fear of global warming rather than fear of the lights going out, but logic has led him to the same destination. Another low-carbon advocate who gets the point about nuclear is James Smith, chairman of the Carbon Trust, a 'not for dividend' organization whose mission is "to accelerate the move to a low carbon economy". Smith is a former chairman of Shell UK, so he has been around the block once or twice. He said, when he retired in 2011 that, internationally, Shell would be more of a gas company than an oil company within two years (which tells us something about the modern energy business), and he was right almost to the day. But it is his pronouncement on nuclear that most concerns us here. Smith recently published a newspaper article headlined "We cannot afford not to have nuclear in our low-carbon energy mix".[141] His gist was that the 'euphoric' phase of low-carbon was now over and there was no solution that was "clean and cheap and always on". There were only three relevant carbon reduction options for the next 20 years: wind, CCS, and nuclear. Wind was free but its availability was uncertain. CCS added 50% to the cost of gas, perpetuated the use of fossil fuel, left us hostage to gas price fluctuations, and had no friends – although Smith believed that if it were to receive the current subsidy for offshore wind, investors would queue up. Nuclear could not ramp up and down as demand fluctuated, but it could deliver huge quantities of low-carbon power.

New nuclear plants were taking longer to build and costing far more than budgeted, Smith acknowledged. To help nuclear over its undeniable cost hump, he appeared to favour the principle of guaranteed pricing for new nuclear, saying "there is a decent case that the consumer takes some of the risk to get the first few nuclear plants going". Consumers should have two serious worries if nuclear priced itself out of the energy market, he concluded. "Firstly, the benefits to innovation and cost control that come from inter-technology competition would be much diminished. Secondly, a key source of diversification of risk to our energy system would be lost. The outcome could be a more costly and risky energy system. Like the other technologies, nuclear has to prove itself in cost terms. But we should not be thinking of giving up now."

A less hysterical approach to radiation and, hence, safety would bring down capital costs. Another development that will do the same is the evolution of small modular reactors (SMRs). Your typical nuclear reactor has a capacity of 1,000MW or more. The new European Pressurised Reactor at Olkiluoto has a nameplate capacity of 1,600MW (1.6GW) and is now thought to be costing €8.5 billion[142] – up from about €3 billion in 2003. That is a terrifying number for an investor who cannot be sure that some new government will not abolish nuclear power the day after the reactor is switched on. SMRs of anything from 10MW to 300MW could make the money less daunting and broaden the market considerably. They are 'modular' because you would be able to install them singly or in series, so you can add capacity as you need it. This would make them suitable for small towns, industrial complexes, and locations remote from large power grids.

Small reactors are nothing new – they have been used in nuclear submarines and other naval vessels for years. Unlike the big ones, each of which must be assembled on site, smaller land-based versions can be 'mass' produced in a factory, with economies of scale that would bring down the cost substantially. Many of the safety features built around larger reactors are more obviously unnecessary in the smaller ones, which will further lower the cost. They are still a work in progress, but the progress is real. In the US, the federal government has been trying to stimulate SMR development, with some success. Among the US groups now developing SMRs of varying technologies and capacities are Babcock & Wilcox (which made components for the world's first nuclear submarine, *USS Nautilus*, so it has a certain amount of experience), NuScale, Holtec International, and Westinghouse (now owned by Toshiba). At the smaller end of the scale, Gen4 Energy is developing a 25MW reactor module that would be sealed, buried underground, and left alone – no refuelling necessary – for all of its 10-year life, at which point it would be dug up and replaced. The idea is that it would compete with expensive off-grid diesel generation, and it could be highly competitive. The company has projected a cost of $50 million per unit, which works out at $2,000 per kilowatt, half that of a traditional (non-Finnish!) nuclear power station.

There is plenty of other SMR development going on, in Argentina, for example, in China, France, India, Russia, and South Korea. Some designs are very small – Toshiba itself has been investigating a 10MW unit that could operate for up to 30 years without refuelling. Others are looking at siting reactors on or under the sea. Rosatom of Russia is developing floating nuclear power stations on giant barges that can be towed to where they are needed. France's Areva is working on FlexBlue, a range of 50MW to 250MW reactors that would be transported by ship, then placed on the seabed and plugged into nearby coastal towns or cities. Like reactors buried underground, this would add improved safety to scalable, affordable convenience. Another virtue of small reactors is that they could be licensed more quickly, which would also help the bottom line. That would be helpful in a jurisdiction such as the US where the laborious process can take 10 years. Some believe that, for small reactors, it could be shortened to two years.[143]

I said that the US government was trying to stimulate work on SMRs. More specifically, the US Department of Energy has launched a $452-million programme to fund the development of 300MW or smaller 'plug and play' reactors. Nuclear is vital to our energy future but the problem is that, unless they get some help, private-sector investors will continue in large numbers to avoid it. The risks and costs make it a no-go for speculative investment, and the balance sheets of most utilities, especially in Europe, are in no shape to be taking this kind of gamble. That is especially true if they do not know what price they will be getting for their electricity five or seven years hence. It would be a grotesque irony if, just as the world was acclimatizing to nuclear power as both acceptable and necessary, there was no risk capital available to fund vital new capacity. I have said all along that subsidy has its place, sensibly targeted and for limited time periods. Well, at this moment nuclear needs, and deserves, a helping hand.

Help could take the form of guaranteed electricity prices for a specified period, tax breaks, or loan guarantees to help new projects raise cheap finance. But even if governments were willing – and some are – this is harder than it sounds in the European Union. European law allows governments to subsidize renewable energy sources such as wind, solar, and hydro, but not nuclear. Nuclear may be clean but, as far as Europe is concerned, it is not 'renewable'. At least four EU member states have been pushing for nuclear, as a carbon-free energy source, to be accorded the same status as renewables and thereby qualify for legitimate subsidy. The Czech Republic, France, Poland, and the UK have all made formal requests for equal treatment for nuclear, although these have been robustly opposed by Germany, which seems determined to impose its own nuclear neuroses on the rest of Europe. It may not have helped that the European Energy Commissioner, Günther Oettinger, is both German and 'cautious' on nuclear power. That said, the European Commission has subsequently agreed to consider changes to the rules, acknowledging that the desire on the part of certain member states to extend support "to other low-carbon energy sources including nuclear merits an in-depth discussion".[144] It is of critical importance that the Commission relaxes the rules, although I am not yet betting the Porsche on the outcome.

To continue feeding all of our energy needs while reducing carbon emissions is going to be a stretch. Mobilizing the huge potential of civil nuclear energy will be essential if we are to meet the challenge. Succeeding in this is so important to the orderly wellbeing of our societies – think New York blackouts – that, if nuclear needs public support, it deserves to get it. The same cannot be said of renewables. No matter how much public money we spend on wind and solar, they will not materially help us to satisfy our energy requirements. And they are not, as we have seen, the *sine qua non* of emission reduction. If the public purse can afford to support nuclear at the same time as renewables, the wind and solar industries would be very grateful. (Shell got out of renewables when it realized it was not the energy business but the manufacturing business.) But if it is a zero sum game, and we can only afford to subsidize one or the other, then nuclear has by far the superior strategic claim.

Peter Atherton, the former Citigroup analyst we met in Chapter 7, pointed out that, as things stand, the risk/reward balance of nuclear for private-sector investors is hugely negative. "But," he said, "nuclear power provides two important public goods – low-carbon base generation, and security of supply. These public goods have no monetary value to private sector investors. So the public must pay for them – only then can new nuclear work for private investors."[145] The big hump is capital expenditure and interest expense, which make up the vast majority of the cost. After that, the running costs are low. So public support for the capex can bring us economically sustainable green baseload – which is exactly what we want.

Wind and solar are not going to wither away. They will take their rightful place in our energy cocktail, when and where they can be competitive or where people are prepared to pay a premium for them. The idea of a vast and wholly renewable European and Mediterranean electricity grid fed by giant CSP farms in the Sahara is intriguing, although I am not holding my breath, and it would be rash to base all of our planning on experimental technology.

If we do not get serious about nuclear and gas, we can be sure of one thing – there will be more carbon dioxide blanketing the planet in the years to come. Already we have the bizarre situation in which the US, non-believers, are reducing carbon emissions while the zealous Europeans are boosting them. Why? Because the pragmatic Americans are using more cheap domestic gas whereas Europe, with expensive gas but not enough cheap nuclear to fall back on, is falling back on cheap coal. And on the course it is setting itself, it can expect more of the same, with Germany already in line for an increase in coal-fired generation because of its irrational aversion to nuclear.

Energy planning, by its very nature, is a long-term affair. Unfortunately, it is at the mercy of politicians, for whom no light illuminates any landscape beyond the next election. They will pander to the popular taste for superficially green policies, if they think that is what will win votes, because in the long-term they will be gone. Sadly, the legacies of ill-conceived policies are more enduring than their political architects.

Concern about climate change is perfectly rational and legitimate. Let us do everything we can to keep our planet clean and habitable. But there is a boundary between sense and silliness. We have to resist the kind of political overkill that endows anything 'green' with a religious aura – a you-cannot-argue-with-this quality that tramples all over any standards of what is practical and sustainable. And when you add free money to the Cult of Green, you get a Bubble, a sudden inflation of interest and activity, borrowing, and investment that would otherwise be more productively occupied elsewhere. Even as I write this, I have spent the afternoon working on the refinancing of a heavily subsidized biogas plant. Without the subsidy, the business would not be viable and its undoubtedly talented entrepreneurs would be getting on with a project that was more sustainable in the long term.

It is, ultimately, as taxpayers, our money that is being spent and, as long as we are reducing emissions, we want the biggest bang for our buck that we can get. I will tell you what is going to happen if we keep pumping liquidity into projects that are not commercially sustainable. It is spookily reminiscent of the last Bubble and the way it ended, and the Bubble before that. For the last few years, we have been blessed with unusually low interest rates. If interest rates are 0.5% and inflation is running at 2%, you are actually being paid to borrow money. With interest rates this low, almost any project can make sense, but it cannot last. Interest rates will head north again, and that will squeeze the green energy sector. Unless they have borrowed at a fixed rate, it will cost them more to service their loans. But their revenue streams, being regulated, will stay the same and the pressure on their margins will build up. We saw this happen in California during the 1990s. Some of these players are very leveraged, borrowed to the hilt, and when there is not enough income to support their borrowing, their whole edifice will collapse.

At some point the politicians will wake up to the fact that it is not working, that they have been pouring our money into businesses that do not have a future on their own and which do not further public (and, therefore, their) interests in any meaningful way. That may be before or, more likely, after the media wakes up to the fact that the emperor has no clothes on and starts to attack wind and solar subsidy. Government will start dreaming up creative ways to withdraw the subsidies in a way that does not technically break their promises but nonetheless pulls the rug out from heavily geared, unsustainable energy businesses. Many of these businesses will have been able to borrow more on a loan-to-value basis than their bankers would have tolerated if they had not been green. They will start to go belly-up and their investors will lose their equity. Their lenders will not see their loans repaid.

In the meantime, the bankers will have been up to their old tricks, repackaging the loans on the back of government guarantees, ignoring whether or not the underlying businesses make commercial sense. We have already heard talk of "sunshine bonds". The bursting of the Bubble will force the weakest of these securitized products into default, hurting their unfortunate investors. Insurance companies, which have been giving special treatment to insurance applications for green projects, will also feel the pain as the trauma spreads. Financial markets are jumpy enough as it is and the damage from a crisis in the subsidized green sector could ripple far and wide.

We have already paid a high price for this fixation with renewables, in terms of higher electricity prices and the diversion of our taxes from other, more immediately productive ventures. But this could be just the start of the cost, if it blinds us to what must be done to keep the lights on between now and mid-century. Add to that the possible disruption of already fragile financial markets and it seems like a high price indeed. Off with the green spectacles. It is time to see the world as it really is. Tell our politicians, on the doorstep or via the ballot box, exactly what they have to do to guarantee our energy future. As physicist Homi Jehangir Bhabha, the father of India's nuclear programme, once said: "No power is costlier than No Power."

REFERENCES

AF-Mercados UK (2012), *Powerful Targets: Exploring the relative cost of meeting decarbonisation and renewables targets in the British power sector.* Edinburgh: AF-Mercados UK.

Allison, Wade (2012), Why radiation from civil nuclear plants should not be treated as exceptionally dangerous. In: Taylor, Corin, ed. 2012. *Britain's Nuclear Future.* London: Institute of Directors.

Birkett, Derek (2010), *When Will the Lights Go Out?* London: Stacey International.

Buchan, David (2012), 'The *Energiewende* – Germany's Gamble'. Oxford: Oxford Institute for Energy Studies.

Crane, Hewitt D., Kinderman, Edwin M., Malhotra, Ripudaman (2010), *A Cubic Mile of Oil.* New York: Oxford University Press.

Department of Energy and Climate Change (2011), *Digest of United Kingdom Energy Statistics.* London: Her Majesty's Stationery Office.

Ernst & Young (2009), *Securing the UK's energy future – seizing the investment opportunity.* London: Ernst & Young.

Etherington, John (2009), *The Wind Farm Scam.* London: Stacey International.

Eurelectric (2012), *Powering Investments: Challenges for the liberalised electricity sector.* Brussels: Eurelectric.

Farrell, J (2012), *Hawaiian Sunblock.* Minneapolis: Institute for Local Self-Reliance.

Frondel, M., Ritter, N., Schmidt, C.M. and Vance, C. (2009), *Economic Impacts from the Promotion of Renewable Energy Technologies – The German Experience.* Essen: RWI.

Greenacre, P., Gross, R., Heptonstall, P. (2010) *Great Expectations: The cost of offshore wind in UK waters.* London: UK Energy Research Centre.

Helm, Dieter (2012), *The Carbon Crunch.* New Haven and London: Yale University Press.

International Energy Agency (2011a), World Energy Outlook 2011. Paris: IEA.

International Energy Agency (2012a), Medium-Term Coal Market Report 2012 Factsheet. http://www.iea.org/newsroomandevents/news/2012/december/name,34467,en.html

International Energy Agency (2011b), Assumed costs for World Energy Outlook 2011. http://www.worldenergyoutlook.org/weomodel/investmentcosts/

International Energy Agency (2011cI), CO2 Emissions from Fuel Combustion – Highlights, 2011 edition. Paris: IEA.

International Energy Agency (2012b), World Energy Outlook 2011 – Executive summary. Paris: IEA.

Gaultier, Donald L et al (2009), Assessment of Undiscovered Oil and Gas in the Arctic. *Science* 29 May 2009. http://www.sciencemag.org/content/324/5931/1175.abstract#aff-1

MacKay, David J.C. (2009), *Sustainable Energy – Without the Hot Air.* Cambridge: UIT.

Martin, Richard (2012), *Superfuel.* New York: Palgrave Macmillan.

Moomaw, W., P. Burgherr, G. Heath, M. Lenzen, J. Nyboer, A. Verbruggen, 2011: Annex II: Methodology. In IPCC Special Report on Renewable Energy Sources and Climate Change Mitigation [O. Edenhofer, R. Pichs-Madruga, Y. Sokona, K. Seyboth, P. Matschoss, S. Kadner, T. Zwickel, P. Eickemeier, G. Hansen, S. Schlömer, C. von Stechow (eds)], Cambridge University Press, Cambridge, United Kingdom and New York, NY, USA.

Mott MacDonald (2010), UK Electricity Generation Costs Update. Brighton: Mott MacDonald.

Pimentel, David and Patzek, Tad W (2005), 'Ethanol Production Using Corn, Switchgrass, and Wood; Biodiesel Production Using Soybean and Sunflower'. Natural Resources Research, Vol 14. No 1.

Massachusetts Institute of Technology (2006), *The Future of Geothermal Energy.* http://mitei.mit.edu/publications/reports-studies/future-geothermal-energy

Schiellerup, Pernille (2011), Energy saving is the key to EU energy and climate goals. In: Barisch, Katinka, ed. 2011. *Green, Safe, Cheap: Where next for EU energy policy?* London: Centre for European Reform.

Taylor, Corin, ed. 2012. *Britain's Nuclear Future.* London: Institute of Directors.

Ürge-Vorsatz, Diana et al. (2010), *Employment Impacts of a Large-Scale Deep Building Energy Retrofit Programme in Hungary.* Budapest: Centre for Climate Change and Sustainable Energy Policy.

ENDNOTES

1   Crane et al (2010) p.22
2   International Energy Agency (2011a) p.74
3   International Energy Agency (2012a)
4   http://www.fossil.energy.gov/programs/powersystems/
    pollutioncontrols/Retrofitting_Existing_Plants.html
5   http://www.pennenergy.com/articles/pennenergy/2013/08/endesa-to-appeal-
    decision-against-740-mw-coal-power-project.htmlhttp://www.pennenergy.
    com/articles/pennenergy/2013/08/endesa-to-appeal-decision-against-740-
6   International Energy Agency (2011b)
7   International Energy Agency (2012a)
8   http://www.guardian.co.uk/environment/2012/
    nov/20/coal-plants-world-resources-institute
9   Gaultier et al (2009)
10 International Energy Agency (2011c) p.8
11 U.S. Carbon Output to Fall to 1970s Levels by
    2040: Exxon, Bloomberg, March 13 2013
12 http://www.aga.org/Newsroom/factsheets/Pages/NaturalGasFacts.aspx
13 International Energy Agency (2012b) p.1
14 International Energy Agency (2011b)
15 World Nuclear Association, http://www.world-nuclear.
    org/info/Nuclear-Fuel-Cycle/Uranium-Resources/Military-
    Warheads-as-a-Source-of-Nuclear-Fuel/#.UUxpzRycQrg
16 International Energy Agency (2011b)
17 International Energy Agency (2011b)
18 Delingpole, James (2012), *Watermelons*. London: Biteback Publishing
19 http://epp.eurostat.ec.europa.eu/cache/ITY_PUBLIC/8-
    10032014-AP/EN/8-10032014-AP-EN.PDF
20 Wall Street Journal Europe, 'EU Emissions Program
    Hit as Prices Drop', 17 February 2013
21 Global Carbon Project, Carbon Budget 2012, www.
    globalcarbonproject.org/carbonbudget/index.htm
22 Database of State Incentives for Renewable Energy, http://www.
    dsireusa.org/documents/summarymaps/RPS_map.pdf
23 European Photovoltaic Industry Association, http://www.epia.
    org/index.php?eID=tx_nawsecuredl&u=0&file=/uploads/tx_
    epiapressreleases/130211_PR_EPIA_Market_Report_2012_FINAL_01.
    pdf&t=1364058901&hash=b828b593fd52ab555fd4fce7a2d3d347194f0b70
24 International Energy Agency, Concentrating Solar Power
    Roadmap, http://www.iea.org/publications/freepublications/
    publication/csp_roadmap_foldout.pdf
25 International Energy Agency (2011b)
26 http://www.globaldata.com/PressReleaseDetails.
    aspx?PRID=156&Type=Industry&Title=Alternative+Energy

27 International Energy Agency (2011b)
28 Massachusetts Institute of Technology (2006) p.1-3
29 International Energy Agency (2011b)
30 UK Department of Energy and Climate Change, *Digest of UK Energy Statistics* 2014, London, 2014
31 http://www.ft.com/cms/s/0/9abb2028-a4ca-11e1-9a94-00144feabdc0.html#axzz2OIAWhG5N
32 http://www.ft.com/cms/s/0/a214ee6a-0155-11de-8f6e-000077b07658.html#axzz2OIAWhG5N
33 International Energy Agency-Energy Technology System Analysis Programme and International Renewable Energy Agency briefing paper (2010), *Marine Energy*
34 International Energy Agency-Energy Technology System Analysis Programme and International Renewable Energy Agency briefing paper (2010), *Marine Energy*
35 http://www.fao.org/forestry/energy/en/
36 International Energy Agency (2011a) p.74
37 Cocchi, Maurizio (2011). Global Wood Pellet Industry Market and Trade Study. IEA Bioenergy Task 40
38 International Energy Agency-Energy Technology System Analysis Programme and International Renewable Energy Agency briefing paper (2012), *Production of Liquid Biofuels*
39 http://www.iea.org/aboutus/faqs/transport/
40 http://www.nytimes.com/2012/07/31/opinion/corn-for-food-not-fuel.html?_r=0
41 Pimentel and Patzek (2005)
42 http://web.mit.edu/aeroastro/people/waitz/publications/AAarticle.pdf
43 http://www.iata.org/SiteCollectionDocuments/AviationClimateChange_PathwayTo2020_email.pdf
44 http://www.economist.com/blogs/babbage/2012/04/electric-cars
45 http://www.chargemasterplc.com/index.php/press-recent-press-cuttings
46 http://www.navigantresearch.com/research/electric-vehicle-market-forecasts
47 Job figures cited in Sandström, Christian (2013), *Are green jobs promising the moon?* Stockholm: Timbro
48 Buck, Stuart and Yandle, Bruce (2001), 'Bootleggers, Baptists, and the Global Warming Battle'. Available at SSRN: http://ssrn.com/abstract=279914 or http://dx.doi.org/10.2139/ssrn.279914
49 http://www.energyandcapital.com/articles/warren-buffett-renewable-energy-investment/2033
50 Frondel (2009)
51 Germans Cough Up for Solar Subsidies, Spiegel Online April 7 2012
52 Grid Instability has Industry Scrambling for Solutions, Spiegel Online August 16 2012
53 Ernst & Young (2009)
54 2009, p.84
55 Alvarez, Garcia Calzada et al., Study of the effects on employment of public aid to renewable energy sources, King Juan Carlos University, 2010

56 UK Department of Energy and Climate Change,
    Electricity Generation Costs, October 2012
57 Farrell (2012)
58 Farrell (2012) and http://www.seia.org/state-solar-policy/hawaii
59 MacKay (2009) p.186
60 Helm (2012) p.79
61 Birkett (2010) p.19
62 *The Dawning of the Digital Energy Revolution*, November 2012
63 'We'll Need Conventional Power Plants until 2050'
    – *Der Spiegel* Online, November 15 2012
64 MacKay, p.33
65 Greenacre et al., p.vii
66 Alvarez, Garcia Calzada et al., Study of the effects on employment of public
    aid to renewable energy sources, King Juan Carlos University, 2010
67 'Revealed: true cost of wind farms.' *The Sunday Telegraph*, June 16 2013
68 Green-Weiskel, Lucia: *China's Green Leap Backward*.
    The Nation, 4 February 2013.
69 http://www.chinafaqs.org/issue/coal-electricity
70 Hall, C.; Cleveland, C. Petroleum drilling and production in the United
    States: Yield per effort and net energy analysis. *Science* **1981**, *211*, 576-579
71 Pimentel and Patzek (2005)
72 http://www.smartplanet.com/blog/energy-futurist/
    why-baseload-power-is-doomed/445
73 The Smart Grid Business 2012 to 2017 (2013)
74 http://www.nrgexpert.com/top-facts-smart-grid-technology/
75 http://www.bloomberg.com/news/2013-02-25/silicon-
    valley-shifting-to-power-grid-after-solar-sours.html
76 Schiellerup (2011), p.53
77 Schiellerup (2011), p.56
78 'Green homes jump in value', Money section, Sunday Times, June 16 2013
79 Wilhite, Harold (2007). *Will efficient technologies save the world?*
    A call for new thinking on the ways that end-use technologies affect
    energy using practices. ECEEE 2007 Summer Study
80 Garnett, T. (2007), *Food refrigeration: What is the contribution to*
    greenhouse gas emissions and how might emissions be reduced? A working
    paper produced as part of the Food Climate Research Network,
    Centre for Environmental Strategy, University of Surrey, England
81 http://energyindemand.com/2013/01/02/increased-self-
    generation-of-electricity-in-german-industry/
82 http://www.smartplanet.com/blog/bulletin/ikea-
    pursues-energy-independence-by-2020/3521
83 European Commission (2011). Green paper - *Lighting the future-*
    accelerating the deployment of innovative lighting technologies
84 https://technology.ihs.com/512707/led-nobel-prize-invention-
    leads-to-177-billion-market-and-250000-jobs
85 http://www.reuters.com/article/2013/02/08/us-
    china-led-idUSBRE91701H20130208

86   McKinsey & Co (2012). *Lighting the way:*
      Perspectives on the global lighting market
87   http://www.designrecycleinc.com/led%20comp%20chart.html
88   European Commission (2013). Energy challenges and policy:
      Commission contribution to the European Council of 22 May 2013
89   Energy priorities for Europe: Presentation of J.M. Barroso, President of
      the European Commission, to the European Council of 22 May 2013
90   Eurelectric (2012)
91   CERA (2012), European Policy Dialogue 2012.
      Cited in Eurelectric (2012), p.12
92   "We have a problem with politics", European Energy Review, 25 February 2013
93   Eurelectric (2012), p.15
94   http://www.future-es.com/utility-finance-in-the-
      2010s-a-lecture-by-peter-atherton/
95   http://www.future-es.com/utility-finance-in-the-
      2010s-a-lecture-by-peter-atherton/
96   Unwrapping British energy bills to 2020,
      Bloomberg New Energy Finance, 2013
97   Ofgem chief warns on energy prices. ft.com, 19 February 2013
98   http://about.bnef.com/press-releases/new-investment-
      in-clean-energy-fell-11-in-2012-2/
99   http://about.bnef.com/press-releases/solar-surge-drives-
      record-clean-energy-investment-in-2011/
100   http://about.bnef.com/press-releases/new-investment-
      in-clean-energy-fell-11-in-2012-2/
101   Renewables 2012 Global Status Report, Ren 21, 2012
102   http://uk.reuters.com/article/2013/03/22/uk-bosch-
      solar-idUKBRE92L0VR20130322
103   Buchan (2012) p.4
104   A Capital Error?: Germany Created Own Threat with
      Chinese Solar Aid, Spiegel Online, 27 February 2012
105   http://english.caijing.com.cn/2013-01-04/112407030.html
106   www.knowledgeatwharton.com.cn/index.cfm?fa=viewArticle&articleID=2424
107   Green-Weiskel, Lucia, China's Green Leap
      Backward, The Nation, 4 February 2013
108   Bloomberg, Vestas Says Chinese Market Shifting to
      Higher Quality Turbines, September 14 2012
109   http://www.thechinatimes.com/online/2012/03/2685.html
110   Annual Energy Outlook 2012
111   World Wind Energy Association, 2013
112   http://online.wsj.com/article/SB125193815050081615.html
113   http://www.reuters.com/article/2013/02/14/us-spain-
      renewables-idUSBRE91D1A020130214
114   http://praguemonitor.com/2013/05/09/zeman-solar-
      power-support-biggest-robbery-history-state
115   http://www.epia.org/fileadmin/user_upload/Press_Releases/Background_
      document_with_negative_measures_in_EU_Members_states.pdf

116 Hughes, Gordon, The Performance of Wind Farms in the United Kingdom and Denmark, Renewable Energy Foundation, 2012

117 Eurostat, H2 2011

118 http://www.ft.com/cms/s/0/745a9e6a-6fca-11e2-8785-00144feab49a.html#axzz2esgDIUl6

119 http://www.nytimes.com/2011/05/04/world/04population.html

120 http://uk.reuters.com/article/2013/04/09/us-africa-summit-population-idUKBRE9380DH20130409

121 IEA (2012b), p71

122 Crane et al (2010), p13

123 http://www.shell.com/global/future-energy/scenarios/2050.html

124 Crane et al. (2010), p270

125 AF-Mercados UK (2012)

126 http://www.ibtimes.com/dangers-relying-hydroelectric-power-brazils-lesson-1056722

127 http://uk.reuters.com/article/2013/06/03/uk-igas-idUKBRE95206H20130603

128 The Economist, July 14 2012

129 Crane et al. (2010), p274

130 Mott MacDonald (2010), p.70

131 Moomaw et al. (2011), p.982 – results for 50th percentile

132 Cited in MacKay (2009), p.168

133 http://www.radiationandreason.com/

134 Allison (2012), p.27

135 Martin (2012), p.73

136 http://www.independent.co.uk/voices/commentators/james-lovelock-nuclear-power-is-the-only-green-solution-6169341.html

137 http://www.guardian.co.uk/environment/2013/feb/19/cannot-afford-nuclear-power-energy-mix

138 http://www.telegraph.co.uk/finance/comment/ambroseevans_pritchard/9784044/China-blazes-trail-for-clean-nuclear-power-from-thorium.html

139 http://yle.fi/uutiset/tvo_unperturbed_by_nuclear_reactors_spiralling_cost_estimates/6415992

140 http://ec.europa.eu/competition/state_aid/legislation/environmental_aid_issues_paper_en.pdf

141 http://www.future-es.com/utility-finance-in-the-2010s-a-lecture-by-peter-atherton/p. 26

INTRODUCTION
TO **PER WIMMER**

Per Wimmer is a Danish philanthropist, space advocate, entrepreneur, financier, adventurer, author and future private astronaut.

Per owns and runs his own investment bank, in London, Wimmer Financial which he founded on the 50th anniversary of Sputnik day, Oct 4, 2007. His bank specializes in global corporate finance within the area of natural resources as well as real estate and infrastructure financing.

Per also runs Wimmer Family Office, an asset management company with investment strategies for wealthy families and high net worth individuals.

WimmerSpace includes Per's three missions to space on three different rockets. Per is a Founding Astronaut with Sir Richard Branson's Virgin Galactic and, thus, one of the first astronauts to travel on SpaceShipTwo. Per is also astronaut number one to travel on the xcor lynx rocket. Wimmerspace is also very active in inspiring children to live out their dreams through various educational and charity programs.

In 1998, he graduated from Harvard University, USA, as an MPA graduate with concentrations in business, finance and international relations whilst receiving the Don K. Price Award for academic excellence, community contributions and potential leadership. He also has a Master In Political Science from College Of Europe, Belgium, a Master Of International Law, LLM from University Of London, UK and a Master Of Law from University Of Copenhagen, Denmark.